Burma Medical Dept

The Burma Medical Manual

Containing Rules for the Management of Charitable Hospitals and...

Burma Medical Dept

The Burma Medical Manual
Containing Rules for the Management of Charitable Hospitals and...

ISBN/EAN: 9783337041649

Printed in Europe, USA, Canada, Australia, Japan

Cover: Foto ©berggeist007 / pixelio.de

More available books at **www.hansebooks.com**

THE BURMA MEDICAL MANUAL

CONTAINING RULES FOR

THE MANAGEMENT OF CHARITABLE HOSPITALS AND DISPENSARIES

AND FOR

THE GUIDANCE OF MEDICAL OFFICERS

UNDER

THE GOVERNMENT OF BURMA.

Issued Under Authority.

RANGOON:

PRINTED BY THE SUPERINTENDENT, GOVERNMENT PRINTING, BURMA.

1898.

[Price,—Rs. 1-6-0.]

PREFATORY NOTE.

This Manual is intended to be the first step towards the compilation of a Medical Code for the province, and all concerned or interested are invited to forward to the Administrative Medical Officer suggestions for its improvement.

In all matters concerning hospitals and dispensaries this Manual, which has been compiled under the orders of the local Government, will be the sole authority and should alone be quoted in official correspondence. Circulars of the Indian Medical Department, Government letters and circulars, and other orders on which the rules in this Manual are based should not be cited, except when the subject to which they refer is not treated of in this Manual.

Correction and addition slips will be issued periodically. These should be pasted into their appropriate places in the Manual, and a note showing that this has been done should be made on the fly-leaves provided for this purpose at the end of the volume.

A copy of this Manual will be issued to each medical institution brought under the supervision of Government or Local Fund Officers, and each subordinate Medical Officer should on receiving charge satisfy himself that the Manual has been kept corrected up to date.

Medical Officers and subordinates may obtain copies for their own use on paying the price of the Manual and the annual subscription for correction and addition slips into the nearest treasury, and forwarding the *chellan* to the Superintendent, Government Printing, Burma.

D. SINCLAIR, M.B., *Surg.-Col.*,

Administrative Medical Officer,

Burma.

CONTENTS.

THE BURMA MEDICAL MANUAL

CONTAINING RULES FOR

THE MANAGEMENT OF CHARITABLE HOSPITALS AND DISPENSARIES

AND FOR

THE GUIDANCE OF MEDICAL OFFICERS

UNDER

THE GOVERNMENT OF BURMA.

CLASSIFICATION.

1. Civil hospitals and dispensaries are divided into the following classes :—

CLASSES I STATE—PUBLIC AND II STATE—SPECIAL —These include all institutions maintained by provincial funds and under Government management. The fact that an institution possesses endowments or receives contributions from local funds or private subscriptions should not be regarded as a reason for not classing it as "State," so long as provincial and imperial funds are practically responsible for all the charges connected with it. Class I.—"Public" are State dispensaries which are open to the poorer classes of the public. Class II are State dispensaries which serve only a special section of the public as indicated in the sub-classification given below :—

    (i) Police.
    (ii) Forests and Surveys.
    (iii) Canals.
    (iv) Others.

CLASS III LOCAL FUND.—Local fund dispensaries include all institutions which are vested in local boards or municipalities or guaranteed or maintained by local or municipal funds. The fact that such an institution is aided by private subscriptions, or receives assistance from Government in the shape of part of the salary of the medical officer, grants of medicine, or otherwise, should not be regarded as a reason for not classing it as a local fund dispensary so long as its existence is practically dependent upon local funds.

CLASS IV PRIVATE AIDED.—Comprises institutions supported by private subscriptions or guarantee, but receiving aid from Government or local funds.

CLASS V PRIVATE NON-AIDED.—Comprises institutions maintained entirely at the cost of private individuals or associations. The fact that Government supplies superior inspection or registers should not be regarded as a reason for not treating it is a private non-aided dispensary.

CLASS VI RAILWAY.—Comprises all railway dispensaries whether maintained by State railways or others.

NOTE.—1. Transfers of dispensaries from one class to another should be reported to the Administrative Medical Officer.
    2. Hospitals of the Dufferin Fund should be shown as private aided or non-aided, as the case may be.
    3. Dispensaries of a purely itinerant character should in no case be included in the classification.

which has accommodation for in-patients, a *dispensary* one for providing out-door relief only

ed by the same authority. Proposals for opening or closing such institutions shall be made in consultation with the Civil Surgeon of the district and transmitted through the Administrative Medical Officer.

3. No institution may be included in Class ~~III~~ unless it receives aid in some way from Government, *i.e.*, unless it is supervised or inspected by Government or local fund officers or receives grants of forms and registers. When a private institution is thus brought into Class ~~III~~, it must conform to the rules in this Manual.

4. Applications for grants from Government in respect of any hospital or dispensary shall be transmitted to Government through the Civil Surgeon, Deputy Commissioner, Commissioner, and Administrative Medical Officer in succession.

The Lieutenant-Governor reserves the right to withdraw Government supervision or aid in any case or at any time when it may seem desirable to do so.

5. No grant may be made by any local authority in aid of any hospital or dispensary which has not received the recognition of Government and been classified under these rules. Grants-in-aid shall be made only in accordance with these rules.

6. A guarantee bond (*see* page xlvii, Appendix I) for the maintenance of a dispensary for a given period may, if thought necessary, be required from the intending supporters of an institution under Class ~~III~~, as a condition precedent to their obtaining assistance from Government in money or in any other way.

7. The committee of every hospital or dispensary organized under these rules, whether in Class I, II ~~or~~ III, shall submit to the Administrative Medical Officer, through the Civil Surgeon of the district, such reports, returns, and accounts as may be prescribed by him from time to time. Such reports, returns, and accounts shall be in such forms and shall be submitted on such dates as may be prescribed. All books, forms, and registers which may be necessary for the preparation and submission of such reports, returns, and accounts will be supplied by Government free of cost on application.

8. Every institution which is recognized by Government and brought under these rules shall be subject to inspection and supervision by the official visitors appointed under these rules. No charge shall be made for such inspection or supervision.

9. The grant of Government aid to a hospital or dispensary shall be conditional on the observance of due economy in the management of the funds. The Administrative Medical Officer shall bring to the notice of Government any instance in which he considers that such economy is not observed and make recommendations as to the continuance or withdrawal of such aid.

10. Every hospital or dispensary under these rules shall admit for examination and treatment all cases brought by the police, and shall provide facilities for the conduct of *post-mortem* examinations.

HOSPITAL AND DISPENSARY COMMITTEES, THEIR CONSTITUTION, DUTIES, &C.

11. Hospitals and dispensaries under these rules shall be managed as follows :—

(*a*) Dispensaries and hospitals included under Class I shall be managed by committees composed of the *ex-officio* members specified in rule 14, and of such persons as the Commissioner of the division may appoint to be members. At outlying places, where a committee would be useless, the Commissioner may, in consultation with the Administrative Medical Officer, appoint any person or persons for the management of these institutions.

(*b*) Dispensaries and hospitals included under Class III. may be either municipal or district cess fund institutions.

The former are managed by a sub-committee of the municipal committee, to which some independent members, selected and appointed by the municipal committee, may be added.

When a municipality pays the entire cost of the dispensary or hospital, there shall be no Government nominees.

When a municipal hospital or dispensary receives a grant-in-aid from Government, the Commissioner of the division shall decide the number of Government nominees on the sub-committee and shall appoint such members.

In small municipalities, where eligible persons are not available outside the municipal body, no non-municipal members need be added to the committee.

District cess fund institutions are managed on the same lines as those in Class I.

(c) Dispensaries and hospitals included under Class III are managed by committees nominated by the supporters and approved by the Commissioner of the division; provided that no committee shall be appointed to any institution in this class which receives no grant-in-aid from Government or local funds, unless the supporters specially apply for such appointment.

12. The names of members appointed by the Commissioner shall be published by the Commissioner in the *Burma Gazette.*

13. On a vacancy occurring in a committee, it shall be filled immediately. Vacancies occurring among the Government nominees of a committee shall be filled temporarily by appointment of the Deputy Commissioner subject to confirmation by the Commissioner of the division.

14. The undermentioned officers shall be the *ex-officio* members referred to in Rule 11, and shall perform the duties herein assigned. This rule does not apply to municipal hospitals and dispensaries.

(a) The Deputy Commissioner shall be *ex-officio* President of all hospital and dispensary committees in his district.

(b) Subdivisional Officers and Township Officers shall be *ex-officio* Vice-Presidents and members of such committees in their respective subdivisions and townships. At stations where there is a Treasury Officer, he also shall be an *ex-officio* member of the committee.

(c) The Civil Surgeon or Subordinate Medical Officer in charge of a hospital or dispensary or both shall be *ex-officio* members of its committee.

15. A member of each committee shall be elected as Secretary of the committee.

16. The Deputy Commissioner shall whenever possible preside at the meetings of such committees in his district; in his absence his place shall be taken by the Subdivisional Officer; in the Subdivisional Officer's absence the Township Officer shall preside, and in the absence of all three any member elected at the meeting shall preside.

17. The duties of the Secretary shall be—

(*a*) To examine, and sign in token of having done so, all establishment and contingent bills and acquittance-rolls.

(*b*) To examine and sign the monthly accounts prepared by the Subordinate Medical Officer prior to their being placed before the dispensary committee, and he shall be held responsible for their correctness.

(*c*) To prepare, in conjunction with the Subordinate Medical Officer in charge, and submit budget estimates of receipts (other than subscriptions) and expenditure of the hospital or dispensary.

(*d*) To give notice of meetings and of the business to be transacted at them, to prepare the proceedings of meetings, and to see that they are duly recorded.

18. Hospital and dispensary committees shall meet monthly and shall scrutinize the monthly accounts and pass them after disallowing any charges considered improper. (But payment of ordinary charges shall not be deferred till after the meeting.)

19. The committee shall inquire into all matters connected with the welfare of the hospital or dispensary; and on such matters as relate to the economy of the institution shall communicate with the Civil Surgeon, who shall act in consultation with the Deputy Commissioner; but in matters of a professional nature (such as the treatment of the sick, departmental returns, food and medical supplies, &c.) they shall act on the advice of the Civil Surgeon and, if necessary, obtain the opinion of the Administrative Medical Officer.

20. The committee shall supervise the expenditure of the funds sanctioned for the maintenance of the hospital or dispensary, and shall sanction all expenditure subject to the control of the Deputy Commissioner, the President of the municipal or town committee, or the Administrative Medical Officer, as the case may be. If the budget provision is found insufficient, the committee shall apply to the Controlling Officer for a reappropriation of funds, giving full reasons for the necessity of the application.

21. The committee shall at every meeting examine the subscription account books and satisfy themselves that the cash in hand is correctly shown. They shall record in the

minutes of the meeting whether the accounts and cash balance are correct and the vouchers complete.

22. It shall be the duty of the committee to inquire into all unfavourable reports and bring to the notice of the Civil Surgeon any misconduct or laxity on the part of the Subordinate Medical Officer in charge, or any defect in the management.

23. The committee shall give special attention to the maintenance of the buildings in a good state of repair, and shall see that their general sanitation as well as that of their surroundings is properly attended to.

24. Every member of the committee is expected to take an active personal interest in the welfare of the dispensary, and to do all in his power to promote its interests and extend its sphere of usefulness.

25. The proceedings of every committee meeting shall be recorded in a minute book, and shall be signed by the President or other member who presides at the meeting.

26. The minutes so signed shall be read at the next meeting before any other business is considered, and shall then be confirmed by the committee, or, if necessary, amended.

27. A copy of the minutes of each meeting shall be sent to the Deputy Commissioner for transmission to the Commissioner of the division, and a second copy shall be forwarded through the Civil Surgeon to the Administrative Medical Officer. These copies shall be sent within a week from the date of the meeting.

28. At every monthly meeting of the committee the following books and statements shall be placed before the members :—

> (1) The visitors' book.
> (2) The subscription book.
> (3) The cash-book.
> (4) The Post Office Savings Bank pass-book.
> (5) The monthly return of income and expenditure.
> (6) The monthly return of receipts and expenditure of subscriptions with all necessary vouchers.
> (7) A statement of patients treated during the month.

VISITORS.

29. The following officers shall be the official visitors of hospitals and dispensaries within their jurisdictions :—

> (a) Commissioners of divisions.
> (b) Deputy Commissioners.

(c) Subdivisional Officers.

(d) District Superintendents of Police.

30. The hospital or dispensary committee shall at each meeting appoint one of their members to be the visitor for the month to visit the dispensary at frequent intervals and see that the institution is being properly managed and that the subordinates are attentive to their duties.

31. Visitors shall record for the information of the committee the results of their visits in the visitors' book. Each visitor shall enter his or her name and the date of the visit in the visitors' book, whether he or she has any remarks to make or not.

SUPERINTENDENTS.

32. The Civil Surgeon shall be *ex-officio* Superintendent of all hospitals and dispensaries in his district classified under these rules.

33. He shall exercise a general supervision over these institutions and shall interest himself in their working. He shall, in fact, be responsible for their good management.

34. He shall periodically visit the various institutions in his district and muster the staff attached to them. He shall at these inspections satisfy himself as to the efficiency of the staff, and see that patients receive every attention and are well cared for; that the instruments and other property of the dispensary are kept in good order and are properly stored; that medicines are sufficient in quantity and of the kinds which are needed and that they are properly expended; that the books and records are correctly written up and arranged; that medical cases are fully and accurately recorded; and that correct treatment is afforded.

35. He shall give Subordinate Medical Officers all needful advice and encouragement, and shall urge them to maintain and improve their professional attainments, and see that clinical aids are fully used by them.

36. The Civil Surgeon shall exercise complete professional control over Subordinate Medical Officers and see that they perform their duties zealously and efficiently.

37. He shall use his influence in stimulating subscriptions and shall carefully scrutinize the expenditure and accounts calling the attention of the committee to any irregularity or other circumstance which, in his opinion, deserves notice.

38. All pay and contingent bills shall be countersigned by the Civil Surgeon before payment.

39. All reports, returns, and accounts required by the Administrative Medical Officer shall be countersigned and transmitted by the Civil Surgeon.

Appointment, pay, leave, &c., of Subordinate Medical Officers.

40. The Subordinate Medical Officer of every institution in Class I of these rules shall be a Government servant and shall be appointed by the Administrative Medical Officer.

41. The Subordinate Medical Officer of every institution in Class II of these rules shall, unless otherwise specially provided by Government, be a Government servant. In cases where special provision is made for the entertainment of an officer other than a Government servant, the appointment shall be only made with the approval of the Administrative Medical Officer.

42. The Subordinate Medical Officer of a dispensary or hospital in Class III of these rules shall ordinarily be appointed by the supporters or committee of such an institution in any manner they may think fit ; but if the institution receives pecuniary aid from provincial funds Government may reserve the power of controlling the appointment of an officer to the subordinate medical charge of the dispensary.

43. The conditions under which an officer other than a Government servant may be nominated under Rules 41 and 42 are—

    (*a*) that the nominee is duly qualified and holds a license, certificate, or diploma from some recognized medical school or college (such license, certificate, or diploma being subject to examination and approval by the Administrative Medical Officer) ;

    (*b*) that the nominee is not a dismissed servant of Government, or disqualified for duty by reason of age, infirmity, or moral unfitness ; and

    (*c*) that in the event of misconduct, insolvency, or professional incompetence (whether by reason of age or otherwise), the Administrative Medical Officer may direct the nominee to be removed from his or her appointment.

44. The Administrative Medical Officer may at any time, for departmental or other reasons, remove or transfer any Government medical subordinate whose services have been lent to an institution included under Classes II and III, and may appoint another in his place.

3, 4, 5

45. The pay to which a Government medical subordinate in charge of any hospital or dispensary is entitled shall be the pay of his grade, and no allowance in excess of the authorized salary* may be granted without the consent of the Administrative Medical Officer and the sanction of Government being first obtained.

46. A Government medical subordinate appointed to the charge of a hospital or dispensary under these rules retains his right to pension and leave allowances under the Civil Service Regulations.

47. All applications for leave from such Government medical subordinates shall be submitted through the hospital or dispensary committee and the Civil Surgeon to the Administrative Medical Officer, who alone is authorized to grant the leave and to appoint a substitute. In cases of illness necessitating early relief, applications for leave from Government medical subordinates may be addressed direct to the Civil Surgeon, who shall transmit them to the Administrative Medical Officer.

48. Subordinate Medical Officers in/charge of hospitals or dispensaries in Class II and Class III of these rules, who are not in Government service, shall be entitled to such leave as the committee may be empowered to grant. The committee shall appoint a *locum tenens* under the provisions of Rules 41, 42, and 43, or shall apply to the Administrative Officer for the services of a substitute, provided that they are prepared to meet the pay of such an officer from funds at their disposal.

49. The salaries or part salaries of Government medical subordinates appointed to the charge of dispensaries of Class II or III, including such enhancements as may become due on their promotion, and contributions (if required) towards their leave and pension allowances, shall be paid by the committee.

50. In special cases where, on a full consideration of the income and expenditure of the funds from which a hospital or dispensary is chiefly supported, the Deputy Commissioner and Commissioner within whose jurisdictions the institution is situated, are satisfied that the funds are really unable to meet the increased charge arising from the promotion of a Subordinate Medical Officer, the Government will be prepared

---

* For authorized scale of salary *see* rule 146, page 27 of Appointments and Allowances Manual (1895 edition).

to consider the expediency of affording assistance either by appointing an officer on lower pay, or by a grant-in-aid equal to the amount due.

Duties of Subordinate Medical Officers.

51. The Medical Officer in charge of a hospital or dispensary shall be responsible for the treatment of the sick, and this duty shall on no account be delegated to a subordinate; he shall also be responsible for all matters connected with the comfort and well-being of the patients, and for the maintenance of discipline.

52. The Medical Officer in charge of a hospital or dispensary shall always be present during the hours appointed for affording medical relief, and shall treat all cases that may be presented.

53. He shall at all times give his immediate attention to urgent cases, and, when not himself at the dispensary, he shall leave the compounder or other competent person in charge with instructions as to where he is to be found if wanted.

54. He is permitted to engage in private practice (with the consent of the committee), but this privilege must not be allowed to interfere with the proper performance of his public duties. He must not absent himself from the dispensary during the appointed hours of attendance, except in urgent cases when the patient or patients requiring his services cannot be moved.

55. Any person attending at a hospital or dispensary is entitled to receive medical advice and medicines free of charge; but the Medical Officer in charge shall impress upon all but the poorer class of patients the duty of subscribing to the funds of the institution. Repeated applications from such persons for gratuitous medical relief shall be brought to the notice of the committee for such action as they may think fit.

56. The Medical Officer in charge of a hospital or dispensary is not compelled to attend patients (whether subscribers or otherwise) at their own houses, except those Government servants who are entitled to such attendance, and he may only in urgent cases visit the latter during the regular dispensary hours.

57. He shall not be required to attend subscribers gratuitously at their own homes.

58. The Medical Officer in charge (except at institutions where a clerk is entertained) shall keep all the books, records,

and accounts of the hospital or dispensary ; he shall also prepare and submit punctually all returns, reports and other documents prescribed in this Manual or by the Administrative Medical Officer.

59.   At hospitals or dispensaries where a clerk is employed, all the clerical work shall be performed under the direction of the Subordinate Medical Officer ; and the latter shall be responsible for the correctness and prompt submission of reports and returns.

60.   The Medical Officer in charge shall be responsible that all medicines and instruments and other property belonging to the hospital or dispensary are carefully and tidily kept.

61.   Every Subordinate Medical Officer shall strive to maintain his hospital or dispensary in the highest state of efficiency, and shall attend to all details in the management, and preserve the buildings and their surroundings in a sanitary condition.   He shall pay particular attention to the conservancy arrangements, and shall see to the proper housing of patients and the segregation of those suffering from any disease of a contagious nature.

62.   An important part of a Subordinate Medical Officer's duties shall be to do all in his power to render the institution popular and to extend its sphere of usefulness among the surrounding population.

63.   It shall be the duty of the Subordinate Medical Officer in charge of a hospital or dispensary at a station where there is no superior Medical Officer present, to conduct all *post-mortem* or other medical examinations and to preserve notes for the information of the Civil Surgeon, Magistrate, or Police Officer as prescribed in rules 212 to 241 and 251 to 259.

ESTABLISHMENTS.

64.   The selection and appointment of menials shall rest with the Subordinate Medical Officer subject to the approval of the hospital or dispensary committee ; but in the case of compounders the appointment shall be subject to the confirmation of the Civil Surgeon, who shall have the power of removing and replacing an inefficient man.

65.   All appointments, removals, and dismissals of servants shall be reported to the Secretary with a view to the matter being brought before the committee for approval at their next meeting.

66. In all hospitals and dispensaries one of the subordinates must always be present and sleep on the premises.

67. No hospital or dispensary servant shall be employed in the private service of Medical Officers.

68. No Subordinate Medical Officer, or compounder, or other servant of a hospital or dispensary shall have any interest in a private dispensary or druggist's shop. A Subordinate Medical Officer may, with the consent of the Civil Surgeon and the hospital or dispensary committee, keep a private stock of medicines for the use of his private patients.

BUILDINGS AND FURNITURE.

69. All buildings in which it is proposed to locate, whether temporarily or permanently, a hospital or dispensary and all alterations and extensions of such buildings shall receive the approval of the Administrative Medical Officer. All plans and estimates shall be transmitted for this purpose through the Civil Surgeon, the Deputy Commissioner, and the Commissioner of the division, who shall, in token of their approval, countersign such plans and estimates, or record any suggestions they may deem fit.

70. Every hospital and dispensary must be provided with quarters for the Subordinate Medical Officer, the compounder, and servants, roofed latrines, a mortuary, and, where inpatients are received, a kitchen and a shed for the accommodation of persons suffering from infectious ailments.

71. Unless more suitable arrangements are available, the dry-earth system of conservancy shall be practised in all hospitals and dispensaries.

72. Every dispensary shall be provided with the following furniture :—

> A writing table ;
> A dispensing table ;
> An almirah for medicines and instruments where proper shelves and cupboards have not been provided ;
> An almirah, box, or drawer for poisons ;
> Two or more chairs ;
> A couple of wooden benches for the use of patients ;
> A wash-hand basin ; and
> A zinc bucket for the mortuary.

At hospitals there shall be, in addition to the above,—

> Cots according to the number of patients capable of being housed ;

Blankets, two at least for each bed;
Pillows, one at least for each bed;
Pillow-cases, two to each pillow;
Sheets, two to each bed;
Bedside tables, one for each cot;
Common earthenware pans or *gumlahs* for the use of
    patients, in sufficient number;
Frames for bed-head tickets, one for each bed;
An almirah, rack, or other suitable receptacle for hos-
    pital linen, clothing, &c.;
A dooly or dandy;
A lamp for the operation room;
A hand lantern;
A medicine tray.

MEDICAL AND SURGICAL EQUIPMENT.

i.—*Indents.*

73. All hospitals and dispensaries classified under Class I,
of these rules shall obtain their supplies of medicines, instruments, and appliances from the Government Medical Stores.

74. Hospitals and dispensaries classified under Classes II,
and III of these rules may make their own arrangements
as to the source from which they obtain their medical stores;
provided that the Administrative Medical Officer is satisfied
that the arrangements are economical.

75. Indents for medicines and surgical equipment from
hospitals and dispensaries specified under Classes II and III
of these rules shall be accompanied by a certificate to the
effect that the cost of the stores indented for does not exceed
the allotment for this purpose in the budget (*vide* page xxix,
Appendix G). The amount of the allotment available shall
be distinctly stated on the certificate and its correctness
attested by the Secretary of the committee. If the institution is maintained by the district cess fund, the certificate
and statement shall be countersigned by the Deputy Commissioner in token of check of budget provision.

76. Annual indents for medical stores shall be prepared
by the Subordinate Medical Officer in charge of each hospital
or dispensary, and shall be submitted in duplicate through the
Civil Surgeon of the district and the Deputy Commissioner
or President of the hospital committee to the Administrative
Medical Officer for countersignature on such dates as may be
prescribed from time to time by that officer.

77. Supplementary and emergent indents shall only be
submitted in cases of urgent necessity, and a memorandum

shall accompany each such indent explaining the causes under which the demand has arisen. Such indents shall be submitted in duplicate with a certificate similar to that prescribed for annual indents, and shall be forwarded through the same channel to the Administrative Medical Officer.

78. Indents for medical stores shall ordinarily be confined to the medicines, instruments, and appliances named in the lists given in Appendix C; and Civil Surgeons shall restrain Subordinate Medical Officers from indenting for all or nearly all the articles on the lists, especially in cases of drugs the action and uses of which are nearly the same. Any exceptional requirement shall be explained in a memorandum forwarded with the indent.

79. In all indents quantities of fluids shall be specified by measure; dry articles, mineral acids, chloroform, essential oils, and carbolic acid shall be entered in avoirdupois weight. These entries shall be carefully scrutinized by the Civil Surgeon; special care shall be taken that the entries in the column of requirements *plus* the quantities "Remaining in store" do not exceed the previous year's expenditure. Every demand in excess of this allowance shall be explained in the memorandum accompanying the indent.

80. The Civil Surgeon shall, whenever possible, test the correctness of the entries in an indent under the column "Remaining in store" by having from 15 to 20 *per cent.* of the stock taken at random weighed and measured in his presence. In cases where errors are detected, the Civil Surgeon shall recommend the Subordinate Medical Officer at fault for exemplary punishment.

81. Extraordinary demands for medicines not entered in Appendix C shall be purchased from a private chemist up to a limit of Rs. 5 for each purchase and up to an annual limit of Rs. 50. This expenditure shall be subsequently recovered on a contingent bill under the countersignature of the Administrative Medical Officer.

82. If in any particular instance it is found impracticable to restrict the purchase of medicines to the annual limit of Rs. 50, a reference shall be made to the Administrative Medical Officer with a full report of the circumstances under which the necessity has arisen, and the Administrative Medical Officer will, if he approves, provide or recommend that provision be made, as the case may be, for an increased grant.

83. As the charges for packing and transit of consignments of sulphuric, nitric, and hydrochloric acids are the same for small and large quantities, and as the medicinal properties of these acids are not impaired by keeping, Medical Officers shall, for the sake of economy and convenience, indent for not less than half a pound at one time, and, when larger quantities are required, for multiples of that amount. Other drugs shall be indented for in the quantities specified in Appendix C.

84. *Spongio impermeable* and *spongio piline* shall not be indented for unless special reasons are given for their necessity. *Spongio impermeable* is not issued from the Government Medical Stores.

85. Bazaar medicines and indigenous drugs of undoubted repute shall be used as much as possible in dispensary practice; and Medical Officers should encourage the use of these drugs as substitutes for the more expensive articles of the British Pharmacopœia.

Explanation.—By the restrictions laid down in the foregoing rules it is not intended to restrict Medical Officers to any particular lines of treatment or to hamper them in their professional practice by unreasonably limiting them in the use of drugs, but Medical Officers are enjoined to afford hearty co-operation in carrying out every measure which tends to secure economy, so far as this is consistent with the performance of all in their power for every patient coming under their care.

ii.—*Medicines.*

86. Medicines shall be stored carefully; and each bottle, vessel, or packet shall bear a label indicating distinctly the name of the contents. Wherever possible, medical stores shall be kept under lock and key, portions of this stock being withdrawn in smaller bottles or vessels from time to time for dispensing prescriptions.

87. No charge shall be made for medicines dispensed to patients who attend at a hospital or dispensary or to Government servants who send prescriptions signed by Government Medical Officers.

88. No medicines shall be issued from the stock of the hospital or dispensary to any one who does not receive treatment either as an in-patient or out-patient. But a Government servant, attended at his house by a Government Medical Officer is entitled to have the prescriptions of that officer made up at the hospital or dispensary, the name of the Government servant being in such cases registered as an out-patient.

89. Unless specially sanctioned by competent medical authority, *i.e.*, by the Civil Surgeon or Administrative Medical Officer, medicines shall not be issued in bulk to patients, but merely in accordance with prescription.

90. No drug shall be issued on the requisition of non-professional persons.

91. Prescriptions prepared by Staff Surgeons for civilian patients residing within cantonments shall be dispensed at a military hospital, and those from Civil Surgeons for Military Officers in military employment residing in the civil lines, at the civil dispensary; provided that the medicines dispensed at a civil dispensary or hospital are not purchased from local funds.

iii.—*Poisons.*

92. All poisons shall be kept under lock and key in a box, drawer, or almirah, and the key retained by the Subordinate Medical Officer.

93. All bottles or vessels containing any of the drugs named in the list of poisons given in Appendix E, shall be labelled "Poison."

94. Poisons shall be kept separate from all other drugs, and on no account shall any poison be placed among other medicines, even though it has been labelled or its nature otherwise indicated, nor shall any non-poisonous drug be placed in the poison box, drawer, or almirah.

95. The poison box, drawer, or almirah shall always be kept locked, and the key shall not be entrusted to the care of a compounder or other hospital servant. Poisons shall invariably be removed as required and replaced by the Subordinate Medical Officer.

96. A list of poisons as given in Appendix E shall be hung near the poison box, drawer, or almirah. Printed lists of poisons are obtainable on indent from the Superintendent, Government Printing, Rangoon.

iv.—*Surgical Instruments and Appliances.*

97. The minimum equipment of surgical instruments and appliances which shall be ordinarily maintained at a hospital or dispensary is given in Appendix D, and duplicates should be kept of such articles only as are in frequent use and are likely to be wanted while the originals are away for repair.

98. The instruments and appliances provided by Government on the opening of a hospital or dispensary are a free gift and become the absolute property of the institution.

99. All instruments and appliances shall be frequently examined to ensure perfect cleanliness. The following instructions shall be observed :—

Steel instruments after use shall be carefully washed and wiped dry with a soft cloth. They shall then be placed in the sun or exposed to an artificial heat of about 140° F. in any manner most convenient for an hour or so. Care shall be taken not to expose the instrument to artificial heat too suddenly as the handles, if made of wood, bone, or other friable material, are likely to be damaged. An excellent way of avoiding this is to lay the instrument on a sheet of tin suspended over a lighted lamp.

When thoroughly dry and while still warm the steel part of the instrument should be plunged into Rangoon oil; the superfluous oil should be allowed to drain off before replacing the instrument in its case or place. In doing so care should be taken not to touch the steel parts of the instrument with the fingers. Instruments made wholly of steel should be handled with a forceps.

Wooden instruments, boxes, and leather cases shall, during the wet season, be occasionally aired in the sun.

Gutta-percha tissues and drainage tubing shall be kept soft and pliable by being completely immersed in cold water, which should be changed at frequent intervals.

Oiled silk shall be kept in an open tray, freely exposed to the air, in a cool place. The sheets shall be first dusted on both sides with very finely powdered sifted chalk, which should be slightly rubbed on.

Gum-elastic articles, such as catheters, stomach-pump tubes, &c., shall be hung up in the air in a cool place and not kept shut up in boxes.

All *aseptic surgical dressings* of lint, cotton, tow, &c., shall be carefully preserved in clean paper wrappers in airtight drawers or boxes.

100. A large case of *post-mortem* instruments will be issued from the Government Medical Stores to institutions where Medical Officers and subordinates are required by Government to conduct such examinations.

101. Measure-glasses shall not be used for administering medicines to patients.

102. The Subordinate Medical Officer in charge of a hospital or dispensary shall submit annually, on the 1st April, through the Civil Surgeon of the district, to the Administrative Medical Officer, a return of surgical instruments and medical

3

and surgical appliances belonging to the institution in form Medl.—35 Misc.

v.—*Unserviceable and Repairable Stores.*

103. Appliances expended in the treatment of the sick, or broken in hospital practice, such as tow, lint, tape, glass, crockery, &c., shall be shown in the column of issues of the return of surgical equipment, and a note will be appended in the column of remarks explaining all cases of breakage or loss.

104. Immediately after the close of the month in which any article of the surgical or medical equipment of a hospital or dispensary is lost, broken, or otherwise rendered unserviceable, a report, in duplicate, in Form Medl.—26 Misc. shall be submitted by the Subordinate Medical Officer through the Civil Surgeon, who shall record on the statement his opinion of the explanation given as to the manner in which an article is stated to have been lost, broken, or otherwise rendered unserviceable, and make recommendations as to whether the article may be written off or whether the whole or part of the cost of the article shall be recovered from the party at fault. The approximate value of the article should be stated.

105. Unless otherwise directed, the amount payable on such account shall be recovered, in the case of hospital subordinates or servants, from the first pay disbursed after the decision of the Administrative Medical Officer has been received. Amounts so recovered shall be credited to the fund from which the hospital or dispensary is mainly supported, and the voucher for such payment placed before the committee at the first meeting after the payment. This voucher shall be sent to the Civil Surgeon if no committee has been organized for the management of the hospital or dispensary.

106. Should the article be required to be replaced at once, a supplementary indent, in duplicate, in Form Medl.—33B. Miscl. shall be submitted through the Civil Surgeon to the Administrative Medical Officer, and the number and date of the decision of the Administrative Medical Officer shall be quoted as the authority under which the indent is submitted; but such indents should be rare as a sufficient reserve of articles in frequent use will be provided at each institution.

107. If not required immediately, and if the article is a necessary part of the equipment of the hospital or dispensary, it shall be included in the annual indent, and the authority quoted as in the preceding rule.

108. All articles rendered unserviceable by fair wear and tear shall be placed before the Civil Surgeon at his inspection of the hospital or dispensary, and a report as laid down in Rule 105 shall be made to the Administrative Medical Officer. The statement with the Administrative Medical Officer's decision shall be placed before the first meeting of the committee convened thereafter, or before the Civil Surgeon at places where no committee has been formed, who shall direct that the articles be destroyed, sent for repair, or sold by public auction, as the case may require.

109. The money realized by the sale of such articles shall be placed to the credit of the fund from which the dispensary or hospital draws its chief support.

110. Amounts realized under Rule 105 or 109 on account of the equipment of hospitals and dispensaries in Class III of these rules shall be received and credited in such manner as the supporters or directors of such institutions may decide.

111. Repairable articles from hospitals and dispensaries in Class I of these rules shall be sent to the Government Medical Stores, and vouchers shall be prepared in triplicate in Form Medl.-34 Miscl.

One of the three copies of the vouchers shall be marked "Receipt" and the other two as "Delivery" vouchers. One copy of the delivery voucher shall be placed in the case containing the repairable articles, and the receipt and remaining delivery voucher shall be forwarded to the address of the Medical Storekeeper.

112. Repairs to instruments, &c., from hospitals and dispensaries in Classes II and III of these rules are executed on payment at the Government Medical Stores, and care should be taken to mark the receipt and delivery vouchers accompanying such articles with the words "On payment" in red ink.

113. Articles supplied from the Government Medical Stores to institutions under Class I may be returned into the stores with the sanction of the Administrative Medical Officer when no longer required. Vouchers as laid down in Rule 111 shall be prepared and marked in red ink with the words "Not to be returned." Hospitals and dispensaries of Classes II and III shall not return such instruments, &c., into the Government Medical Stores.

114. The address on packing cases of articles sent to the Government Medical Stores shall be secured so as not to be easily defaced or torn off, and shall bear the name of the hospital or dispensary from which they are returned.

115. Articles of small value shall not be sent for repair or returned to the Government Medical Stores.

116. Earthenware vessels lost, broken, or otherwise rendered unserviceable shall be disposed of by the Civil Surgeon without reference to the Administrative Medical Officer.

Patients.

117. In-door patients shall be required to conform to all the rules of the hospital and to obey the directions of the Medical Officer in charge.

118. Subordinate Medical Officers in charge of a hospital or dispensary shall discharge any patient who refuses to observe the rules, or to submit to his orders, or to follow the treatment prescribed. The power herein conferred should be exercised with great care and discretion, due consideration being given to the condition of the patient.

119. Each bed when occupied shall be furnished with a bed-head ticket in Form Medl.-3 Miscl. These tickets shall be written up daily, morning and evening, by the Subordinate Medical Officer in charge so as to preserve a concise history of the case and its treatment. It is also advisable to maintain a case-book for recording all special or important cases.

120. Where special accommodation has been provided for paying patients, fees may be levied according to a scale to be sanctioned by the Hospital Committee, or by the Civil Surgeon if no committee has been organized.

121. No charge shall be made for attendance given to an in-patient.

122. In-patients who are permitted to diet themselves shall not introduce any article of food or drink which the Subordinate Medical Officer in charge may consider unwholesome or injurious; and they shall not give their meals to other patients without permission.

123. Friends and relatives may visit the wards of in-door patients at hours to be fixed by the Subordinate Medical Officer in charge. Visits may not be paid at other hours without special permission of the Subordinate Medical Officer in charge, who may, if he thinks fit, ascertain the object of the visit.

124. All patients brought by the police to a hospital or dispensary shall be handed over to the police when discharged, and all particulars in connection with such cases shall be carefully noted in the Register of Police Cases (*see* Procedure relating to Police Cases, page 33).

125. The clothing of patients suffering from an infectious or contagious disease shall be disinfected or burnt.

126. Articles of clothing and other moveable property belonging to a deceased in-patient who has no relatives or heirs shall be forwarded at once to the Magistrate.

127. The funeral expenses of pauper patients shall be borne from the funds provided for contingencies; and the disposal of corpses brought by the police to a hospital shall also be paid for from the contingent grant unless the corpses are examined elsewhere than in the hospital dead-house. In the latter case the expenditure shall be borne by the Deputy Commissioner and charged to the head 32, Miscellaneous— Donations for charitable purposes.

128. Whenever a non-commissioned officer or soldier dies in a hospital, the Subordinate Medical Officer shall inform the Civil Surgeon, who shall report the circumstance to the Administrative Medical Officer.

129. Whenever a Military Officer under the care of a Civil Surgeon is placed on or taken off the sick list, a medical certificate in A. II. Book Form No. 3 shall be furnished by the Civil Surgeon, in the case of an Officer attached to a native corps, to the Medical Officer of the corps, and, in the case of an officer of the British Army, to the Medical Officer in charge of the Station Hospital.

130. Native Military Officers and Inspectors of Police requiring medical treatment shall be attended by the Medical Officer under whose charge they are placed, at their quarters if they are suitable for a sick officer and the illness be of a mild nature; otherwise one of the rooms of the hospital shall be made available for their reception. If there is only one room, a portion shall be screened off so as to afford seclusion. On no account should officers of these classes be relegated to the common ward of a hospital. If it is not possible to provide such accommodation, the officers shall be treated at their own quarters.

131. Out-patients shall attend during the recognized hours. Each patient shall be furnished with a ticket in Form Medl.-8 Miscl., which should be produced each time a renewal of medicine is required, but medical aid shall not be refused if the patient has lost or forgotten to bring this ticket.

ATTENDANCE.

132. The hours of attendance at a hospital or dispensary shall ordinarily be for not less than three hours in the morning, between 6 and 10 A.M., or until all the patients have

been afforded medical relief, and two hours in the evening, between 3 and 6 P.M.; but these hours may be modified by the committee with the concurrence of the Civil Surgeon and Deputy Commissioner.

133. The hours of attendance shall be clearly stated in English and in several native languages on a notice-board posted in some conspicuous place at or near the dispensary or hospital.

134. All Medical Officers and subordinates employed by Government are bound to give their professional services without remuneration when required by proper authority in the interests of Government.

135. Medical attendance in cases of accident or emergency shall be given at all hours, and patients in such cases shall be received unhesitatingly into any hospital or dispensary where admission is sought and where sufficient accommodation is available. Any question as to the particular institution into which a patient might most appropriately be admitted should be deferred.

136. If a patient seeking admission under the above circumstances is suffering from an infectious or communicable disease, he shall, if no infectious diseases shed is provided, be accommodated apart from the other inmates of the hospital, room being made available by clearing one of the wards of less serious cases.

137. No patients except Government servants are entitled to gratuitous medical attendance at their homes.

It shall, however, be generally understood that Government servants of the lower grades will attend at a dispensary or hospital for medical relief when able to do so, and will not demand attendance at their homes for simple ailments.

138. When a Government servant lives at a greater distance than 2 miles from the dispensary, the Medical Officer called on to afford aid shall be provided by him with the means of conveyance.

139. Gratuitous medical attendance shall be afforded to—

(1) all Government servants in civil (*i.e.*, non-military) employment ;

(2) all Royal Indian Marine Officers and Engineers of all grades, when on shore ;

(3) European military pensioners pensioned direct · from the Military Service ;

(4) Sergeant-Instructors of Volunteers ;

(5) All officers in Military employment ;

(6) Officers serving in the Military Works Department;
(7) Officers of the Invalid Battalion;
(8) Cantonment Magistrates;
(9) Chaplains;
(10) Honorary Commissioned and Warrant Officers of the effective and retired lists;
(11) Non-Commissioned Officers in Staff Military employment;
(12) Soldiers transferred to the unattached list;
(13) Civilian employés (*not being clerks*) of the Ordnance Department;
(14) Civilian and Military clerks of Army headquarters and of the office of the Director-General of Ordnance in India;
(15) The families of those named under clause (3) when the pensioner is not in civil employment, and the families of all other officials [except those named in clause (1)] when not in civil employment.

For the purposes of these rules the term *family* includes the Officer's wife and legitimate children residing with and wholly dependent on him, and also his parents, sisters, and minor brothers if wholly dependent on and residing with him.

140. The following officials and their families are, however, not so entitled :—

(i) Unemployed General Officers and Officers transferred to the unemployed supernumerary list or to the retired list, who remain in India for their own convenience.
(ii) Colonels and Lieutenant-Colonels permitted to reside in India.
(iii) Lieutenant-Colonels of the Royal (late Indian) Artillery and Engineers, transferred to the Indian supernumerary list, who are unemployed and not borne on the strength of the garrison of the station wherein they may reside.
(iv) Those who have been granted a pension or gratuity for service in the civil department.

141. The following are entitled to the services of the Civil Surgeon himself :—

(i) Government servants, whether gazetted officers or not, drawing a salary of Rs. 250 and over;

(ii) Officiating Assistant Superintendents of Police;

(iii) Royal Indian Marine Officers and Engineers of all grades when on shore and their families;

(iv) European military pensioners who have been pensioned direct from military service and who reside outside cantonments;

(v) Sergeant-Instructors of Volunteers and their families when there is no Honorary Surgeon of the Corps or Military Surgeon in the station;

(vi) The officials named in clauses (5) to (14) of Rule 139 when there is no Military Surgeon in the station. The families of officers are also entitled to this privilege if the officials are not in civil employment.

142. The attendance of the Civil Surgeon shall, however, be readily given in all cases of emergency or of great danger or of difficulty, or when applied for by the Subordinate Medical Officer.

143. When a Civil Surgeon resides within or near cantonments, the purely Civil establishments in cantonment entitled to his services shall be attended to by the Civil Surgeon; otherwise the Staff Surgeon is the proper medical attendant.

144. When there is no Civil Surgeon at a station, Government servants residing in the immediate vicinity of a cantonment, although not within the actual boundary, are entitled to the services of the Staff Surgeon or other Medical Officer under whose charge they are placed.

145. All officers and the families of officers who are entitled to gratuitous medical attendance in a civil station or cantonment are entitled to it at any other place in India where they may be residing, whether on duty or on leave, from any Medical Officer at the station paid by Government for staff, civil, or general duties.

146. An officer entitled to gratuitous medical attendance is also entitled to gratuitous advice from an officer in the employment of Government called into consultation by the medical officer who is in accordance with regulations in charge of the case. Should the officer, however, decline the attendance of the medical officer provided by Government, and call in another medical officer to him in illness the usual fees can be claimed by the latter.

147. All vessels visiting any port in the province shall have the option of sending their sick to the civil hospital or

dispensary for treatment, and any sailor or other member of a vessel applying for medical relief in person at a sea-port hospital shall be admitted immediately if the case is urgent; otherwise admission may be refused if no written permission from the master of the vessel is produced.

148. On the admission of an urgent case from a vessel in port or of a case where the master of a vessel has authorized admission to a sea-port hospital, the master shall be invited to pay the donation mentioned in Rule 174. If, however, the master refuses to pay, the matter should not be pressed.

149. When a native Chief, or unofficial Gentleman of high position desires the professional attendance of any Medical Officer of Government, the latter will be at liberty to attend him, provided that such attendance does not interfere with the due performance of his ordinary duties. The special permission of the local Government will, however, be necessary when such attendance involves the absence of the officer from those duties for any substantial time.

FEES FOR MEDICAL ATTENDANCE AND CERTIFICATES.

150. To obviate confusion and misunderstanding, the Government of India has ruled that when a Medical Officer is first called in, it is advantageous to all concerned that a clear understanding should be arrived at regarding the fees chargeable, and that, in the absence of a special agreement on the basis of a yearly payment, it shall be fair to assume that a Medical Officer's professional services will be paid for by the visit.

151. When a Medical Officer is called to render professional aid, he shall respond unhesitatingly and leave the question of fees for consideration after his first visit, and may transfer the further treatment of the case to the proper medical attendant.

152. All questions which may arise regarding the amount of remuneration for medical attendance will be left to private adjustment.

153. A Native Chief or Gentleman of high position may offer any Medical Officer of Government attending him such fee as he thinks fit. The offer made will be reported by the Medical Officer through the Political Agent or other officer of Government exercising political functions in the State of which the said Chief, Noble or Gentleman is a resident, for the consideration of the local Government within whose jurisdiction the Native State is situated.

154. In reporting the offer, the Medical Officer will state, so far as he is able to do so consistently with his position as a medical adviser—

(*a*) the nature and extent of the relief afforded ;

(*b*) the importance of the case from a professional point of view ; and

(*c*) the circumstances under which he attended the patient.

155. This report will not be required from a Medical Officer in the following cases :—

(*a*) when the officer names his own charge in accordance with a scale which he has fixed for his patients generally, who are not Native Princes or Chiefs and when such charge has been accepted by the patient ; and

(*b*) when the officer and patient reside in the same station and the fee does not exceed Rs. 50 for each visit, or Rs. 1,000 in the aggregate for repeated visits during the course of a year.

156. When it is necessary to forward a report such as prescribed in Rule 153 the Political Officer shall transmit it with such remarks as he may consider necessary. The local Government will have authority to sanction the acceptance of any fee so reported, unless the amount appears out of proportion to the relief afforded and to the circumstances of the case, in which event the question will be submitted, with the local Government's opinion, for the consideration and orders of the Government of India.

157. The term '*fees*' in these rules includes honoraria or presents from Native Chiefs, Nobles, and Native Gentlemen of high position in a Native State.

158. A Medical Officer, not being the Civil Surgeon or an officer in medical charge of a civil station, shall be entitled to a fee of Rs. 16 for conducting a *post-mortem* examination, and to a fee of Rs. 10 for conducting a *medico-legal* examination other than a *post-mortem* examination, in cases not falling within the ordinary discharge of his duties, whether or not he is required to give evidence in a court of justice in connection with such examination. It is, however, to be distinctly understood that when such an officer is required, under these circumstances, to give evidence in a court of justice, he shall not be entitled to any remuneration in addition to the fee above sanctioned other than the usual expenses paid to a witness.

159. Medical Officers are permitted to charge fees for certificates of physical fitness if the applicant is at the time of examination not in the service of Government, and the certificate is required by any authority other than Government.

160. Civil Surgeons receive a fixed fee of Rs. 4 for each candidate examined under the Postal Life Insurance Rules.

MEDICAL CERTIFICATES.

161. Medical certificates to candidates for employment in Government service shall be in the form prescribed by Article 61 of the Civil Service Regulations.

162. When a Medical Officer finds it necessary to detain an applicant for leave under the European Services Rules, a certificate in the form given under Article 356 (*a*) of the Civil Service Regulations shall be granted.

163. An applicant for an invalid pension, or for leave, or extension, or commutation of leave on medical certificate, shall be provided with a statement of his medical case and of the treatment adopted as required by Articles 487 (*a*), 894 (*a*), and 903 of the Civil Service Regulations, and with the appropriate certificate prescribed by Articles 491 (*a*), 893, and 907.

164. Gazetted officers shall be furnished with three copies of the statement of their medical case and certificate in Form Medl.—13.

165. Unless otherwise ordered, the certificates required by Articles 61, 356 (*a*), 491 (*a*), and 893 shall not be granted by any other than an officer holding the position of a Civil Surgeon.

166. Certificates from a non-gazetted officer's medical attendant shall be in Form Medl.—17 and shall be countersigned by a Civil Surgeon when that officer is not himself the medical attendant.

167. Certificates recommending the grant of casual leave shall be given in Form Medl.—15.

168. No alterations other than those distinctly prescribed by the Civil Service Regulations shall be made in any certificate. Alterations where permissible shall be attested by the initials of the Medical Officer granting the certificate.

169. If a public servant is ill, he must be reported sick. If it be considered that the station in which he is serving is inimical to his constitution, and that he is likely to have better health elsewhere, he is still to be reported sick and unfit for duty and dealt with accordingly. Medical Officers are prohibited from recommending a change of station for a

public servant because the one in which he is serving is not suited to his constitution.

170. When a person is required to be confined in a lunatic asylum, a medical history sheet as appended to Form Medl.—17 L.A., shall be forwarded with the lunatic to the Superintendent of the lunatic asylum. The certificate of the Medical Officer furnished on Form Medl.—17 L.A., is to be filled in only when the persons are sent to the lunatic asylum under section 4, 5, or 8 of Act XXXVI of 1858. (*See* note at foot of Form Medl.—17 L.A., and instructions in Circular No. 98 of 1896 reproduced as Appendix E of this Manual.)

SUBSCRIPTIONS.

171. A subscription book shall be maintained in Form Medl.—16 Misc. at each hospital and dispensary classified under these rules, and subscriptions shall be invited in such manner as may be considered most effectual. Hospital and dispensary committees are recommended to maintain a monthly register of subscriptions and donations in Form Medl.—16 Misc.

172. The names of all persons who subscribe not less than Rs. 10 *per annum* to a hospital or dispensary shall be recorded in English, Burmese, and in such other languages as may be deemed desirable, on a list hung up in a conspicuous position in the building. This list shall be renewed from year to year, the names of those subscribers only who continue their subscriptions being brought forward.

173. Special mention shall be made in the annual report of the hospital or dispensary, for the information of the Administrative Medical Officer, of those persons who have contributed towards a hospital or dispensary in money, buildings, appliances, medicines, or otherwise to the value of Rs. 50 and upwards during the year of report.

174. Any commander of a vessel sending one or more sick persons into a seaport hospital shall be invited to pay an entrance donation, according to tonnage, as noted below :—

			Rs.
Below and up to 300 tons	...	...	15
Above 300 and up to 600 tons	...	...	20
Above 600 and up to 1,000 tons	...	...	25
Above 1,000 tons	...	...	30

The payment of this donation will entitle masters of vessels to send as many seamen as they choose to the hospital during the next twelve months ; and the patients shall be provided with bedding, hospital clothing, furniture, and accommodation suitable to their position free of charge.

175. Money received as subscriptions shall not be paid into the treasury, but shall be paid into the Post Office Savings Bank through the Secretary, or by the Subordinate Medical Officer as collected. Only such portion of sums received as is needed for current expenditure should be retained by the Subordinate Medical Officer.

176. The Subordinate Medical Officer shall observe the rules prescribed in Appendix J regarding the accounts of subscription money.

177. Subscriptions shall, unless expressly otherwise stipulated by the donor or subscriber, be expended in providing the patients with extra comforts, such as iron bedsteads, special nurses, extra diets, and the like.

178. All sums received as subscriptions and the expenditure therefrom shall be accounted for to the committee and the Administrative Medical Officer.

179. The Post Office Savings Bank book shall be retained by the Secretary, or, where no committee has been organized, by the Subordinate Medical Officer, who shall be responsible for the prompt deposit into the Post Office Savings Bank of the sums received.

180. The Subordinate Medical Officer in charge of a hospital or dispensary, at every monthly meeting of the committee, shall place a statement in Form Medl. 16, duly signed by the Secretary, showing the transactions of the subscription fund during the previous month, and shall attach thereto necessary vouchers for all expenditure. A copy of this statement shall be forwarded to the Civil Surgeon, together with the other returns of the month for transmission to the Administrative Medical Officer.

Diet.

181. A diet-scale as given on pages xxvi and xxvii of Appendice G shall be hung up conspicuously in every such hospital.

182. A sheet in Form Medl.—4 Misc. showing the diet and extras which a patient is receiving, shall be placed in the frame holding the bed-head ticket.

183. A register of diets issued shall be kept and shall be written up daily by the Subordinate Medical Officer.

184. Patients may be permitted to diet themselves. Such diet shall be in accordance with that prescribed by the Subordinate Medical Officer in charge of the hospital.

185. On no account shall money be given to a patient to diet himself unless the Subordinate Medical Officer has satisfied himself that the diet is such as will be suitable.

186. When ordering diets, Subordinate Medical Officers will bear in mind the religion of the patients and the restrictions as to food which this imposes on them.

187. Diets should be issued to relatives of patients only in cases where the Medical Officer is satisfied that these people are entirely dependent on the patient, or are unable to earn a livelihood owing to their attention being required by the patient. Such diets should be issued with care, and particular attention should be given to see that the privilege is not abused.

MISCELLANEOUS.

i.—General.

188. A visitors' book shall be maintained at every hospital or dispensary coming under these rules, and visitors shall be invited by the Medical Officer in charge to record any remarks or suggestions they may wish to make in connection with the hospital or dispensary, or its working.

189. Two copies of remarks made by Commissioners and Heads of departments and one copy of remarks by other visitors shall be forwarded to the Administrative Medical Officer through the Civil Surgeon, who shall see that all necessary explanatory notes are given in the margin and a note made of any action that has been taken or is intended to be taken. Where two copies are made, one copy only shall be so annotated ; the other shall be drawn up with a half-page margin, for the remarks of the Administrative Medical Officer.

190. Copies of remarks shall be neatly and correctly made, having a blank half-page margin along the left of the first page and on the right of the second for the notes required from Medical Officers by the above rule.

191. Letters shall invariably be written on paper not smaller in size than a half-sheet of foolscap. A margin shall run along the left side of the first page and the right of the second page to admit of their being stitched into files.

192. Not more than one subject shall be dealt with in each letter ; all enclosures to a letter shall be distinctly specified and securely attached to their covering letters.

193. Bills and returns of various kinds shall not form enclosures to one letter, but each set should be submitted separately.

194. Letters to the same address shall all be placed under one cover, and care should be taken not to affix stamps in excess of the value required by the weight. A saving in postage may be effected by forwarding returns, bills, &c., by book post.

195. Petitions from subordinates shall bear an endorsement by the officers through whom they are transmitted and shall contain such information as will help the officer whose orders are required to dispose of the petition without further reference.

196. When a return is blank, a copy of Form Misc.-Genl. 93, intimating the name of the return, shall be forwarded to the officer concerned.

197. Vernacular communications shall, whenever possible, be accompanied by translations.

198. Printed forms shall be neatly and carefully filled in, and no unauthorized heading shall be added, nor shall any existing headings or ruled lines be erased. Any deviation from this rule should be explained in a forwarding letter. In forms provided with a blank space for remarks, the words " No remarks " should rarely appear.

199. All corrections and alterations, especially in the accounts, books, statements, registers, indents, &c., shall be neatly made in red ink and attested by the President or Vice-President, or, if he is authorized to do so, by the Secretary. Similarly, all alterations and corrections in a voucher shall be duly authenticated by the payee or receiving officer. Erasures should on no account be permitted in registers, statements, vouchers, or accounts of any description.

200. All copies of letters, indents, remarks, &c., shall be true reproductions of the originals and shall be certified as such.

201. The budget estimate of receipts (other than subscriptions) and expenditure of each dispensary shall be prepared by the Subordinate Medical Officer and scrutinized carefully by the Secretary prior to being placed before the committee.

202. The budget estimate shall be framed in time to be placed before the August meeting of the committee, and shall then be forwarded with a copy of the remarks of the committee to the Civil Surgeon. This officer shall, after scrutinizing the estimates, forward them to the Deputy Commissioner, President of the municipal committee, or other controlling officer, with such suggestions as he may deem fit.

203. All payments due to Government servants lent to dispensaries or hospitals shall be made at the same rates and with the same regularity and punctuality as if the servants were in Government employment.

204. The stock and records of every hospital or dispensary under these rules shall be open to the inspection of the Civil Surgeon or Administrative Medical Officer. The accounts of a hospital or dispensary shall be available for the information of all inspecting officers.

205. A list of the moveable property of each hospital and dispensary shall be kept in the following books in the prescribed forms :—

A stock book of all furniture and fittings.

A surgical equipment ledger, Form Medl.-35 Misc.

206. A Subordinate Medical Officer receiving charge of a hospital or dispensary shall check the entries in the above books with the property in the institution and furnish a certificate in Form Medl.-38 Misc. The medicines may be checked in the manner indicated in Rule 80. All discrepancies shall be brought to notice at once by the officer receiving charge. Neglect of these precautions will render him liable for any loss or damage that may have occurred.

207. The records of a hospital or dispensary shall be kept complete up to date, and when a Medical Officer relinquishes a charge, he will satisfy his successor that this has been done, and that the information necessary for compiling all returns has been furnished. The relieving officer will report to the Civil Surgeon any case of non-compliance with this rule.

208. When a transfer of charge takes place a report in Form Medl.-2 Misc. of delivery and receipt of charge shall be forwarded to the Administrative Medical Officer. This report shall in the case of Subordinate Medical Officers be submitted through the Civil Surgeon of the district.

209. When it is not possible to effect a transfer of charge in a day, the date on which the delivery and receipt of charge is completed shall be held to be the date of transfer. The report of delivery and receipt of charge shall therefore not be signed till the transfer proceedings are fully completed, and the officer who is being relieved shall remain in charge till such proceedings are completed.

210. Officers must arrange to take over charge of appointments to which they are posted within the joining time admissible under Chapter IX of the Civil Service Regu-

lations. Extensions of joining time on the ground that the transfer of office occupied more than one day will not be granted.

211. A service-book in Form T. F. No. 33 must be maintained for every subordinate in Government employment. The instructions printed on the fly-leaf of the book should be carefully observed. All service-books must be kept under lock and key in the office of the Civil Surgeon; and when a subordinate is transferred, his service-book should be sent under registered cover to his new departmental superior. (*See* also instructions regarding service-books in the Civil Service Regulations and Burma Treasury Manual.)

ii.—*Procedure relating to " Police Cases."*

212. Whenever a case is sent by the Police for treatment to a hospital or dispensary, it is the duty of the Police authorities to forward at the same time a memorandum of the circumstances of the case in Police Form No. 75. Omission to comply with this rule should be brought at once to their notice by the Medical Officer in charge.

It is the duty of the Medical Officer to examine carefully every police case immediately after arrival at the hospital or dispensary.

213. The Medical Officer shall enter all particulars required by the register of Police cases, Form Medl.—36 Misc. and shall preserve in the same register a detailed history of every case and its treatment from the date of admission to that of discharge. All entries in this register shall be made by the Medical Officer himself.

214. In the event of any patient being in a dangerous state, either at the time of admission or at any subsequent period, an immediate report in Form Medl.—22 Misc. with a view to obtaining a dying declaration, is necessary.

215. If a patient in a police case quits the hospital without the permission of the Medical Officer, his departure shall be reported in writing by the Medical Officer at once to the Police Officer by whom the case was sent to the hospital.

216. The Medical Officer shall bring to the notice of the police authorities any case of serious injury in any case which has not been sent to hospital by the Police unless he shall have reason to believe the cause of the injury to have been accidental.

217. When a patient brought to a hospital by the Police dies or is fit to be discharged, the Medical Officer shall

5

intimate the fact in writing to the Police Officer by whom the case was sent to hospital.

iii.—*Rules for the guidance of Medical Officers in Criminal Investigations.*

218. A Medical Officer shall, immediately on receiving from any person for examination a corpse or any other substance, enquire and note down the name and residence of such person, and, if he be a Police Officer, his number and rank, and shall, without delay, grant to such a person a receipt in Form Medl.—21 Miscl., for the corpse or other substance delivered by him.

219. All corpses and articles forwarded to a Medical Officer for medico-legal purposes shall be examined as soon as practicable after arrival.

220. In cases when a corpse is sent to him, the Medical Officer shall note—

 (i) the time of its arrival ;

 (ii) the date and hour of *post-mortem* examination ;

 (iii) the sex, height, and apparent age of the deceased ;

 (iv) the state of the body, whether well nourished or otherwise ;

 (v) the existence or absence of any caste or other marks not of recent origin. such as cicatrices, deformities, and the like ; and

 (vi) whether the marks on the corpse correspond with those mentioned in the Police report.

221. In cases where he is taken to the place where the body lies, besides the above points, the Medical Officer shall note—

 (i) the place and nature of the soil (if out in the open country) where he finds the body ;

 (ii) its position and the state of the clothes, if any ;

 (iii) the position of the corpse in reference to surrounding objects, such as sharp stones and the like, contact with which, it may be alleged, has produced the injury ; also

 (iv) whether any blood-stains are visible on such objects or anywhere near the corpse ; and

 (v) whether any weapons are lying near it.

222. In the case of a corpse found near or in a pond, river, tank, well or other body of water, the Medical Officer shall note—

 (i) the depth of the water ;

    (ii) any evidence of a struggle having taken place by the margin of the water;

    (iii) if any ligature is found on the body and how it is tied;

    (iv) any mark of violence on the body, head, or neck;

    (v) whether any froth is present in the mouth or nostrils;

    (vi) whether the skin is rough or smooth;

    (vii) whether the hands are clenched. If anything is grasped in the hand, it should be carefully removed, and any mud found under the finger-nails should be compared with the mud at the bottom of the water.

223. In cases of hanging or strangulation the Medical Officer shall note—

    (i) the height of the body from the ground and, before cutting it down—

        (a) whether there is any lividity of the lips or eyelids,

        (b) any protusion of the eyes,

        (c) the state and position of the tongue,

        (d) any evidence of the escape of fluid from the mouth or nostrils, and the direction in which it has flowed;

    (ii) the arrangement and number of knots on the ligature;

    (iii) the state of the clothing, whether torn or dishevelled;

    (iv) on cutting down the body, the state of the neck, particularly the direction of the line of the strangulation, and whether there is any bruising along that line;

    (v) whether the thumbs are crossed on the palms;

    (vi) if possible, the materials by means of which the hanging or strangulation has been effected should be brought away.

224. In cases of suspected poisoning,—

    (a) the Medical Officer shall ascertain—

    (i) whether the person was in ordinary health;

    (ii) whether the symptoms from which the deceased suffered were—

        (a) sudden in onset,

        (b) uniform in character,

(c) rapidly increasing in severity,

(d) speedily fatal;

(iii) what interval elapsed between the last eating and drinking, and (a) the first appearance of symptoms, and (b) death, if it occurred;

(iv) the nature of the first symptoms and whether vomiting or purging occurred;

(v) whether the person became drowsy or fell asleep;

(vi) whether cramps or twitching of the limbs were observed, or tingling of the skin or throat complained of;

(vii) whether any other symptoms were observed;

(b) all food, drink, tobacco, &c., in the house or near the body should be brought away under seal;

(c) if there is any appearance of vomiting having occurred, note nature and position in relation to body;

(d) if vomiting has occurred, the vomit shall all be brought away under seal. Any clothing soiled with vomited matter, and even the flooring into which some may have soaked, should be scraped and brought away;

(e) the condition of the mucous membranes of the mouth, pharynx, and œsophagus, and whether there is any unusual appearance or mark of corrosion thereon should be carefully noted.

225. In every case the Medical Officer shall describe the condition in which he found the body, noting the degree of coldness, warmth, rigidity, and putrefaction, and the amount and nature of the clothing or covering on it.

226. In conducting his examination the Medical Officer should make a careful inspection—

(i) of the bones with a view to determining whether any are fractured or dislocated;

(ii) of the vertebral column, the teeth, the hair, orifices of the body, and general surface;

(iii) of the state of the pupils, whether contracted or otherwise;

(iv) of the neck for marks of compression;

and faithfully record all abnormal conditions.

227. If there be any injuries on the body he should note—

(i) in the case of wounds and contusions, (a) their position, (b) length, (c) breadth;

(ii) the depth and direction of all wounds;

(iii) whether there are any cuts on the clothing corresponding to them;

(iv) whether any foreign bodies are present in the wounds, preserving such as may be found.

228.   He should state clearly—

(i) whether, in his opinion, the wound was mortal;

(ii) whether the wounds could have been self-inflicted; or

(iii) whether they might have been the result of accident;

giving reasons for his opinions.

229.   He should carefully examine any gun, sword, blood-stained instrument, stick, or stone by which the wounds may have been inflicted, and mark such instrument so as to be able to recognize it if asked to do so.

230.   He should also compare the weapon with the wound alleged to have been caused by it, and state whether, in his opinion, it was possible for the wound to have been produced by it.

231.   He should commence his dissection of the body by removing the top of the skull in the usual way with a saw and note anything that appears unusual.

232.   He should then make an incision from the chin down to the pubes, so as to be able to examine the windpipe, heart, lungs, liver, stomach, spleen, kidneys, and intestines, also the urinary bladder, and note whether any of these organs appear diseased, and whether any wound on the outside of the body communicates with the contents of the chest or abdomen.

233.   In making his examination, he should disturb as little as possible any organ which may communicate with an external wound, if he has reason to think that the body may be re-examined by another medical man.

234.   In the case of females, he should examine the ovaries and uterus, bearing in mind that abortion is sometimes caused by the introduction into the uterus of pointed instruments which may cause death. He should note the presence or absence of pregnancy, the probable period to which pregnancy had advanced, and examine the external generative organs for marks of violence.

235.   In the case of infants, he should note the condition of the umbilicus and cord, if any of the latter remain. He

should also remove the lungs and try whether they sink, or nearly sink, or float in water.

236. He should keep all his original notes, even though he may make a fair copy of them afterwards, and should not lend them to any one to read.

237. In all cases the examination of the body should be thorough and the notes of the appearances discovered should be as minute as possible.

238. Full notes should also be made in cases of wounded persons.

239. When summoned to give evidence in any case in which he has made a *post-mortem* examination, or examined a wounded person, the Medical Officer should bring with him to court the original notes made by him at the time of conducting such examination.

240. The notes of examination in all cases, or a fair copy of them in the handwriting of the Medical Officer, should be at once made in the register of *post-mortem* cases (Form Medl.—25 Mis.) kept at the hospital or dispensary for the purpose, and should be signed by him.

241. After having made a *post-mortem* examination in a case of suspected poisoning the Medical Officer should, if he thinks it necessary, report the result thereof to the Police and forward the viscera to the Chemical Examiner to Government for analysis. In cases where no death has occurred, but where it is suspected that poison has been administered, the Medical Officer should report the case to the Police, and should, if he thinks it necessary, forward the vomited matter or the contents of the stomach removed by the stomach-pump to the Chemical Examiner.

242. In all cases of death from suspected poisoning, the stomach should for purposes of analysis be tied at both ends and removed from the body so that its contents may be retained ; after removal, it should be opened, the contents received into a perfectly clean bottle, and the mucous surface of the stomach carefully examined, its appearance noted, and any suspicious particles found adherent thereto should be picked off with a pair of forceps and placed in a separate small phial for transmission.

243. The following articles should be forwarded for analysis, each in a separate bottle, in cases of death from presumed poisoning —

    (*a*) the stomach ;

    (*b*) the contents of the stomach ;

(*c*) suspicious particles (if any have been found) removed from the mucous membrane of the stomach ;

(*d*) a portion of the liver, not less than 16 ounces in weight, or the whole liver if the organ weighs less than 16 ounces, and one kidney. These may be sent in the same bottle ;

(*e*) the vomited matter, if any ; and

(*f*) in cases where it is suspected that death has been caused by administration of *datura* or other vegetable poison, the contents of the small intestines.

(i) Strong rectified spirit or a solution of salt should in all cases be added, as laid down in rule 248, to the contents of bottles (*a*), (*d*), and (*f*), and also to the contents of bottles (*b*) and (*e*), unless it is suspected that alcoholic poisoning has been the cause of death. No spirit need be added to the contents of bottle (*c*).

(ii) All bottles should be carefully sealed by the Medical Officer, and closed in such a manner that they cannot be opened without destroying the seal. The seal used should be the same throughout and should be a private seal ; each bottle should be labelled and the labels should be signed by the Medical Officer.

244. Articles forwarded to the Chemical Examiner for examination should each bear a distinctive mark or number as well as seal, and a letter should be sent at once to the Chemical Examiner advising their despatch. This letter should contain—

(*a*) An impression of the seal used in closing the bottles and a description thereof, together with a copy of the distinctive marks or numbers used ;

(*b*) a list of the articles forwarded and a statement as to how the articles have been forwarded ;

(*c*) a detailed account of the *post-mortem* appearance observed ;

(*d*) if he has seen the case during life, an account of the symptoms observed, and a statement of the treatment, if any, adopted ; and

(*e*) any information which may be furnished to him by the Police or which may come otherwise to his knowledge as to the circumstances which render it necessary that the substance sent should be analyzed. For instance, such information in the cases of supposed death by poisoning would mention the circumstances under which the poison is believed to have been administered, the food or other substance in which it is believed to have been consumed, and so on.

245. When articles are forwarded by post to the Chemical Examiner to Government, each package should be franked externally with the name and address of the Medical Officer forwarding the articles and the rules for the transmission of such articles by post or railway (*see* Rule 248) should be attended to.

246. Parcels containing articles received from the Police should be opened and examined by the Medical Officer receiving them with a view to ascertaining whether any substance is present which requires preserving. Articles should on no account be transmitted to the Chemical Examiner without being subjected to such an examination.

247. When the Medical Officer receives a parcel back from the Chemical Examiner, he should see whether the seals are intact and should note the marks on the parcel and on the articles within it. He should hand over the parcel only to a responsible police officer or to a Magistrate and should take a receipt for it.

248. When articles are forwarded by post or by railway parcel—

(i) The suspected *viscus* or other portion of the body to be sent for examination should be enclosed in a glass bottle or jar, fitted with a stopper or sound cork.

(ii) If liable to decomposition it should be treated in the following way :—

(*a*) In cases of suspected poisoning in man, other than alcoholic poisoning, it should be immersed in spirits-of-wine, which should be used in the proportion of one-third of the bulk of the material.

N.B.—Care should be taken that common bazaar spirit is not used.

(*b*) In cases of suspected alcoholic poisoning in man and in all cases of suspected cattle-poisoning it should be immersed in a solution of common salt. The solution should be prepared by adding common salt to water at the ordinary temperature of the air, stirring all the time until no more salt dissolves, and the solution should then be filtered through a plug of cotton wool. The bottle containing the *riscus* or other portion of the body should be filled up with the solution to within one-quarter of an inch of the stopper.

N.B.—To obviate any danger of the solution of salt being tampered with a separate sample of the solution should in every case be sent to the Chemical Examiner by the officer despatching the *riscus*, and a second sample should be retained in a sealed bottle in his office.

The use of the preservative enjoined in this rule should be invariable, whether the season is hot or cold.

(iii) The stopper or cork should be carefully tied down with bladder or leather and sealed. To ascertain that it has been securely closed, the bottle or jar should be placed for some minutes with its mouth down.

(iv) The glass bottle or jar should then be placed in a strong wooden or tin box, which should be large enough to allow of a layer of raw cotton, at least three-fourths of an inch thick, being put between the bottle or the jar and the box.

(v) The box itself should be encased in common garah cloth, which should be sealed in accordance with the usual rules of the Post Office as to parcels.

(vi) A declaration of contents to the officials of the Postal Department is unnecessary and should not be made.

249. Despatching officers will be held personally responsible that these instructions are carefully followed. Whenever practicable such parcels should be packed under the immediate supervision of the Civil Surgeon.

250. At all stations where there is a Civil Surgeon the parcels should invariably be sent to the post office by that officer and not by a subordinate officer, unless the Civil Surgeon is absent from headquarters, in which case the articles should be packed and forwarded by the senior subordinate Medical Officer of the station. At stations where there is no Civil Surgeon the articles should be packed and forwarded direct to the Chemical Examiner by the medical subordinate in charge of the hospital or dispensary at such station.

iv.—*Rules for the guidance of Medical Officers in cases of suspected cattle-poisoning.*

251. The carcase should be first carefully examined, more especially about the genitals, below the ears, and soft skin of the thighs. If any puncture is found, it is possible that *sui*-poisoning has occurred. The spike or *sui* should then be sought for and, if one be found, it should be wrapped in paper, sealed, and labelled. A chemical examination of the viscera is useless in cases of *sui*-poisoning as in such cases poison cannot be detected in the viscera.

252. The mouth, rectum, and vagina should be examined and anything unusual found in them should be preserved and labelled.

253. The stomach, after removal, should be cut open and emptied. Its appearance (*i.e.*, whether congested or not) should be noted. About one-third or one-fourth (according to its size) should be cut off, selecting the most congested portion, and put into a bottle, and about half a pound of the contents should be put in a separate bottle. A solution of salt should be added in the proportion of one-third of the bulk of the material and the bottles closed, sealed, and despatched as described in Rules 242, 243, 245, 248 to 250. A few ounces of liver may also be sent if it is thought desirable, but no part of any other organ is required. Any suspicious looking leaves, seeds, or plants found in the stomach should also be forwarded in a separate bottle.

254. Dried cattle-dung may be sent without addition of spirit.

255. The form of label should be as follows :—

> *Portion of stomach and contents from a* be-
> *longing to* *sent by A B of D*

> *Dated*

256. Where samples from the stomachs of several animals are sent, they should be consecutively numbered on the labels. The duplicate label in this case may embrace them all thus—

(1) *Portion of stomach and contents from a cow-calf belonging to A B ;*

(2) *Portion of stomach and contents from a bull belonging to C D ;*

(3) *Portion of stomach and contents from a cow belonging to E F ;*

(4) *Portion of stomach and contents from a cow be-longing to G H;*
   sent by **Y Z**, *Civil Surgeon at N. M.*
   Dated

257. Medical Officers shall, on the requisition of a Magistrate or of a Police Officer not lower in rank than an Inspector undertake the *post-mortem* examination of cattle suspected to have died of poison and shall prepare and despatch (in accordance with Rules 242, 243, 245, 248 to 250) to the Chemical Examiner such portions of the viscera as are required for analysis.

258. When a Medical Officer receives articles for despatch to the Chemical Examiner for examination, he should endeavour to obtain from the police full information regarding the symptoms of the suspected poisoning, the reasons for entertaining such suspicion, and such other particulars as would usefully supplement the report furnished by the police. In forwarding the articles he should at the same time address and forward separately a letter to the Chemical Examiner advising their despatch and containing the statement made to him by the police. This letter should contain—

(*a*) an impression of the seal used in closing the vessels and a description thereof ;

(*b*) an exact copy of the label or labels used ;

(*c*) a list of the articles forwarded and information as to how the articles have been forwarded ;

(*d*) the name of the officer from whom the order to forward the articles has been received and the number and date of such order ;

(*e*) information as to the number and kind of animals affected and number of deaths ; and

(*f*) any information obtainable as to *post-mortem* appearances, nature and duration of symptoms, &c., which may be likely to indicate the probable nature of the poison.

259. The following are the poisons usually employed in poisoning cattle in this province :—

White arsenic (8$).
Yellow arsenic (8$ol).
Abrus precatorius (ဂျွ:ဝဃ် or ချိန်ရွှေ). (Hind. *gunchi* or *kunch*).
Seed used for making the '*suis*' *Calotropis gigantea* (ဝမ္မို:). (Hind. *mudar* or *akund*).

Jatropha purgans (သဃဝ်ကြက်ဆူ). (Hind. *ujur jahar*).
Datura alba (ဝတ္ထုင်:ဖြူ or ဝတ္ထုင်:ဝတ္ထာ).
Datura fastuosa (ဝတ္ထုင်:နှ). (Hind. *dhatur*).

v.—*Rules for the Preservation and Destruction of Records.*

260. All hospital and dispensary registers, returns, letters and records of every description shall, for the purposes of preservation or destruction, be arranged so as to render easy the selection of those which have to be destroyed. Medical Officers in charge of district hospitals and dispensaries shall classify correspondence, &c., under headings which facilitate reference, and shall see that the same practice is followed at all subdivisional dispensaries within their respective districts.

261. Documents bearing on or relating to standing orders, rules or regulations, important and permanent charges on the funds by which the institutions are maintained, such as the acquisition and renting of lands, pensions, changes in the sanctioned establishment, &c., shall be arranged under appropriate heads and placed in case-covers prescribed for this purpose (Form Misc.—Genl.-123).

262. The following books shall be maintained in the office of every Civil Surgeon and care shall be taken to keep them corrected :—

   (1) Burma Medical Manual.
   (2) Inspection Manual.
   (3) Examination Manual.
   (4) Excise Manual.
   (5) Civil Account Code, Volume I.
   (6) Burma Account Manual.
   (7) Burma Treasury Manual.
   (8) Burma Travelling Allowance Manual.
   (9) Civil Service Regulations.
 (10) Manual of Appointments and Allowances.
 (11) Burma Laws List.
 (12) Burma Rules Manual.
 (13) Quarterly Civil List.
 (14) Quarterly List of Hospital Assistants.
 (15) History of Services—

      Volume I.—Gazetted Officers.
      Volume II.—Other Officers.

 (16) Service-books of hospital assistants, compounders, and other Government servants attached to hospitals and dispensaries.
 (17) Annual reports on hospitals and dispensaries and on sanitation and vaccination in Burma.
 (18) Circulars of the local Government.

(19) Circulars of the Indian Medical Department.
(20) Hehir and Gribble's Medical Jurisprudence (where already supplied).
(21) Guard-book of forms—
    (*a*) Medical.
    (*b*) Miscellaneous.

263. The following records shall be permanently preserved at all hospitals and dispensaries :—

(1) Register of police cases.
(2) Register of *post-mortem* examinations.
(3) Register of important and interesting cases.
(4) Subscription collection books.
(5) Subscription cash-book.
(6) Stock-book of furniture and fittings.
(7) Acquittance-roll of permanent and extra establishments.
(8) Minute-book of Committee meetings.

264. The following records shall be preserved for a period not less than five years after they have been completed, and then destroyed :—

(1) Register of out-patients.
(2) Register of in-patients.
(3) Register of diets.
(4) Register of surgical operations.
(5) Surgical equipment ledger.
(6) Visitors' book.
(7) Despatch book.
(8) Register of letters received and issued.
(9) Indents for medicines and surgical equipment.
(10) Monthly statistical returns.
(11) Contingent register.
(12) Travelling allowance register.
(13) Budget estimates.

265. The following records shall be preserved for a period of not less than three years and then destroyed :—

(1) Copies of salary bills.
(2) Detailed lists of sanctioned establishment.
(3) Indents for clothing, furniture, &c.
(4) Indents for stationery.
(5) Bills and vouchers in connection with subscription money.
(6) Treasury receipts.

(7) Charge reports.
(8) Nominal disposition returns
(9) Indents for forms.
(10) Annual returns and reports on individual hospitals and dispensaries.
(11) Detailed bill for contingent charges.
(12) Abstract bill for contingent charges.
(13) Register of receipts from paying patients, sales of instruments, &c., other than subscription money.
(14) Breakage statements.
(15) Receipt and delivery vouchers of surgical instruments and appliances.
(16) List of unserviceable and repairable articles.
(17) Statements of property of deceased patients.
(18) Copies of letters requiring deposition to be taken.
(19) Copies of letters forwarding articles for chemical examination.
(20) Copies of medical certificates.

266. The following shall be disposed of in the manner shown against each :—

(1) Bed-head tickets (2) Diet-sheets	To be destroyed two years after discharge of patient to whom they relate.
(3) Contract bonds ...	To be destroyed at end of three years after expiry of contract, or, if the document is registered, after six years.
(4) Transfer receipts... of medical stores, instruments, and appliances.	To be destroyed on receipt of fresh receipt when change of officers takes place.

267. The registers, books, indents, &c., mentioned above shall be maintained in the forms prescribed by the local Government and included in the guard-book of forms. A list of the forms so prescribed and included is given as Appendix H to this Manual, with instructions as to the manner in which they are to be filled in (*vide* Appendix G).

vi.—*Rules for the custody of Forms and Registers.*

268. Indents for forms and registers shall be submitted in Form Mis.—Gen. 102 through the Administrative Medical Officer to the Superintendent, Government Printing, Burma.

269. Except in Rangoon, indents for treasury forms shall be submitted to the Deputy Commissioner of the district.

270. Civil Surgeons will prepare a consolidated indent in triplicate for the hospitals and dispensaries in their dis-

trict and submit it so as to reach the Administrative Medical Officers not later than the 31st May of each year.

271. Duplicate copies of indents will be returned to Civil Surgeons with the forms indented for.

272. No form will be supplied which has not been included in the guard-books.

273. A stock-book of forms and registers shall be kept in Form Mis.—Gen. 103, and the quantities received and the quantities issued from time to time shall invariably be noted therein.

274. As soon as possible after the receipt of forms, &c., from the Superintendent, Government Printing, they should be unpacked and every item carefully checked by the indent. As the contents of packages are checked they should be carefully put away in almirahs, or on racks set apart for this purpose.

275. Records, printed forms, registers, &c., shall be stored in the manner prescribed by the local Government in Financial Department Circular No. 25, dated the 14th November 1890 (page 248 of Local Administration Circulars, 1888—1896).

APPENDICES.

CONTENTS OF APPENDICES.

APPENDIX A.

EXAMINATIONS AND PROMOTIONS.
(i) *Professional Subjects.*

1. Assistant Surgeons in Government service are advanced to a superior grade on their passing the examinations detailed below:—
A written examination in the following subjects:—

Medicine.	Midwifery.
Surgery.	Medical Jurisprudence.

2. The general intelligence and professional knowledge of candidates in each of these subjects will be further tested by a *vivâ voce* examination, which, in the case of the two first-mentioned subjects, namely, medicine and surgery, shall be thoroughly practical and in great part conducted at the bed-side in connection with cases of sickness or surgical injury in hospital.

3. The number of marks required to secure a pass shall not be less than 50 *per cent.* in each subject and each paper.

4. Hospital Assistants in Government service are advanced to a superior grade on their passing a written examination in the following subjects:—

Anatomy.	Materia Medica.
Surgery.	General Sanitation.
Medicine.	Vaccination.

5. Not less than four practical questions will be set in each subject, and at least 50 *per cent.* of the marks obtainable in each subject are required to secure a pass.

6. Assistant Surgeons are required to serve seven years in a grade before they can appear for the examination, but the Administrative Medical Officer may, as a special concession, allow an Assistant Surgeon to appear at the examination held within the half-year immediately preceding the completion of the septennial period.

7. Hospital Assistants are also required to serve seven years in each grade, but may present themselves for examination on the completion of six years' service in each.

8. Promotion in all cases will be granted from the third and second grades to the second and first grades respectively only if subordinates have been well reported on in the Annual Confidential Reports and have qualified in Burmese (*see* paragraph 20, page vi).

9. Examinations will be held in the months of May and November for Assistant Surgeons, and in April and October for Hospital Assistants. These examinations will ordinarily be held on the first Monday and following days of the months named, but these dates are subject to revision.

10. Applications should be made by Medical subordinates desirous of appearing for examination to the Medical or other officers under whom they are immediately employed, who will submit the names of all applicants in Form Medl.—39 Misc. These nominal rolls should

reach the Administrative Medical Officer not later than the dates specified below:—

Assistant Surgeons—15th March and 15th September.
Hospital Assistants—15th February and 15th August.

11. Medical subordinates, both of the Assistant Surgeon and Hospital Assistant classes, who fail to pass in more than two subjects, will be considered to have failed altogether. Those who fail in one or two subjects will be examined again in these subjects only, provided they are re-examined within six months from date of failure; otherwise they will be liable to re-examination in all subjects, and will continue to be so examined until they do pass or until their services are dispensed with under the provisions of clause 12.

12. Subordinates who fail to pass the examination for promotion three times in succession, or who do not succeed in passing within two years of the expiration of each septennial period, shall be removed from the service, unless special extenuating circumstances can be adduced, when they may be given another opportunity.

(ii) English qualification.

13. Hospital Assistants who have qualified by the Seventh Standard Anglo-Vernacular Examination in Burma, or such other examination as may be considered to be equivalent or more than equivalent thereto, are entitled to the higher rate of pay for qualification in English. All others who have not so qualified are required to pass an examination in the following subjects before they become entitled to the higher rate of pay :— .

(a) Ability to read fluently and intelligently ordinary English prose.

(b) A knowledge of orthography and ability to write from dictation with a reasonable amount of correctness.

(c) A complete knowledge of simple arithmetic as far as the rule of three.

(d) Ability to read and write prescriptions in English intelligently.

14. The knowledge of English possessed by Hospital Assistants who qualify under the above test for the enhanced rate of pay will be re-tested at the period of their examination for promotion to the second and first grades respectively, and such subordinates only will be allowed the option of answering the questions set for the promotion examination in the vernacular.

15. Candidates desirous of appearing for the English qualification examination should apply to the officer under whom they are immediately employed, who will forward the application to the Administrative Medical Officer for orders.

16. The re-testing of the knowledge of English at examinations for promotion will, unless otherwise specified, be conducted by the officer superintending the professional examination.

17. The results of the English qualification examination will be reported to the Administrative Medical Officer in Form Medl.—18 Misc., and a certificate in Form Medl.—19 Misc. will be granted to successful candidates by the officer who conducts the examination. A copy of this certificate shall accompany the report of results.

18. Candidates will be considered to have passed in each subject of the English examination if they obtain half or more of the maximum marks fixed for each subject on Form Medl.—18 Misc.

19. When a Hospital Assistant fails only in one or two of the professional subjects at an examination for promotion, but passes in English, he will not be again subjected to an examination in that language if he goes up for the examination in the professional subjects within six months from the date of his failure.

(iii) Burmese.

20. Non-Burman medical subordinates will receive no grade promotion and will obtain no increment of pay until they have passed in the Burmese language by the elementary standard, and on failure to pass the examination within two years of appointment shall be liable to forfeiture of appointment.

21. A reward of Rs. 200 will be paid to Medical subordinates not natives of Burma who pass the lower standard test in the Burmese language.

APPENDIX B.

Misconduct and Punishment of Government Servants.

1. When a Government servant is alleged to have misconducted himself, definite charges, specifying the nature of the offence and the occasion when it was committed, should be framed by the officer under whom he is serving and given to the accused, who should at the same time be called on to furnish in writing, within a reasonable period, any defence he may desire to make with reference thereto, and show cause why he should not be fined, removed or dismissed or otherwise punished.

2. If the alleged offence is of a grave nature, the officer under whom the subordinate is employed is authorized to suspend him at once, submitting a report without delay to the Administrative Medical Officer, detailing the circumstances which necessitate the adoption of this measure and the arrangements made for the performance of the duties of the offender.

3. A copy of the charge or charges, with the defence in original, should be forwarded to the Administrative Medical Officer with such remarks on the defence as may seem called for and a distinct recommendation as to the punishment to be inflicted.

4. "Dismissal" should be recommended only in cases in which the circumstances of disgrace are such as render it desirable that the offender should not be re-employed under Government *in any capacity.*

5. If the circumstances of an offence are such as to necessitate a departmental enquiry where the statements of witnesses or others have to be taken, the witnesses should not be examined on oath nor should advocates be allowed to appear in such cases without the previous permission of the officer authorizing the enquiry.

6. The procedure detailed above is intended to apply specifically to all Government servants employed at hospitals and dispensaries

but reports against members of the menial establishment will be dealt with as laid down in Rule 65 of this Manual. Compounders in Government service may be punished by their immediate superior officers, but such punishments shall be reported to the Civil Surgeon of the district for confirmation and a copy of the proceedings shall be forwarded to the Administrative Medical Officer.

7. Any of the punishments detailed below may be awarded to Medical subordinates of the Hospital Assistant class :—

(a) Reprimand, or extra duty not exceeding 12 hours, to be awarded by the Civil Surgeon or other Medical Officer in charge.

(b) Fines not exceeding five days' pay and allowances in any one month, to be inflicted only on confirmation by the Administrative Medical Officer.

(c) Deprivation of leave, to be awarded by the Civil Surgeon of the district and reported to the Administrative Medical Officer.

(d) Stoppage of promotion to a higher grade for a period not exceeding one year, to be awarded by the Administrative Medical Officer.

(e) Deprivation of approved service allowance or other additional allowances may also be recommended by Civil Surgeons.

8. All punishments will involve entry in the subordinate's statement of service appended to the Annual Confidential Report.

9. A Hospital Assistant reduced to a lower grade may be placed at the bottom of that lower grade, i.e., below all who were in that grade on the date on which such degradation was inflicted.

10. A Hospital Assistant reduced to a lower grade will be eligible for promotion to his former grade on the recommendation of his immediate superior officer after not less than one year from the date on which such reduction was ordered.

11. Re-examination for promotion will be necessary only in the case of a Hospital Assistant who has not passed the test for promotion within two years of the date on which he is recommended for restoration to his former grade.

Table showing the authorized drugs and quantities in which such drugs issue on

[See Rules 78,

NOTE.—Dry articles, mineral acids, chloroform, essential oils, and carbolic acid are sup
fluids by

HALF A POUND IS THE SMALLEST QUANTITY OF CONCEN

Serial No.	—	Lbs.									
	I.—Medicines.										
1	Acaciæ Gummi	8	6	5	4	3	2	1½	1	12	8
2	Acetum Scillæ	...	...	...	...	...	2	1½	1	12	8
3	Acidum Aceticum	...	...	...	4	3	2	1½	1	12	8
4	Acidum Benzoicum	...	...	...	...	...	...	...	...	8	6
5	Acidum Boricum	8	6	5	4	3	2	1½	1	12	8
6	Acidum Carbolicum	2	1½	1	...	...	...	...	...	12	8
7	Acidum Citricum	...	...	...	...	...	2	1½	1	12	8
8	Acidum Gallicum	...	...	...	...	...	2	1½	1	12	8
9	Acidum Hydrobromicum Dil	...	...	...	...	...	...	...	1	12	8
10	Acidum Hydrochloricum	4	3½	3	2½	2	1½	1	½	...	...
11	Acidum Hydrocyanicum Dil	...	...	...	...	...	...	...	...	8	4
12	Acidum Nitricum	4	3½	3	2½	2	1½	1	½	...	...
13	Acidum Salicylicum	...	...	...	...	...	...	...	1	12	8
14	Acidum Sulphuricum	4	3½	3	2½	2	1½	1	½	...	...
15	Acidum Tannicum	...	...	...	...	...	2	1½	1	12	8
16	Acidum Tataricum	...	...	...	...	...	2	1½	1	12	8
17	Alumen	4	3	2	1½	1	...	...	...	12	8
18	Ammonii Carbonas	8	6	5	4	3	2	1½	1	12	8
19	Ammonii Chloridum	4	3	2	1½	1	...	...	...	12	8
20	Antifebrin	...	...	2	1½	1	...	...	...	12	8
21	Antimonium Tartaratum	...	...	...	...	...	...	...	...	8	4
22	Antiseptic Solution	4	3	2	1½	1	...	...	...	12	8
23	Apomorphinæ Hydrochloras	...	...	...	...	...	...	...	...	...	...
24	Aqua Destillata	4	3	2	1½	1	...	...	...	...	8
25 {	Argenti Nitras	1	...	...	...	...	...	...	...	12	10
	Argenti Nitras toughened	...	...	...	...	...	...	...	...	12	10
26	Asafœtida	1	...	...	...	...	...	...	...	12	8
27	Atropinæ Sulphas	...	...	...	...	...	...	...	...	..	...
28	Bismuthi Subnitras	4	3	2	1½	1	...	...	...	12	10
29	Borax	2	1	...	...	...	...	...	...	12	8
30	Buchu Folia	2	1	...	...	...	...	...	...	12	8
31	Camphora	8	6	5	4	3	2	1½	1	12	8
32	Camphorodyne	2	1	...	...	...	...	...	...	12	8
33	Cerii Oxalas	...	...	...	...	...	...	...	...	4	3
34	Chirata	8	6	5	4	3	2	1½	1	12	8
35	Chloral Hydras	2	1½	1	...	...	...	...	...	12	10
36	Chloroformum	3	2	1½	1	...	...	...	...	12	10
37	Chrysarobinum	1	...	...	...	...	...	...	...	12	8
	Cinchona Alkaloids, namely,—										
38	(a) Cinchona febrifuge	1½	1	...	...	...	...	...	...	12	8
39	(b) Cinchonidinæ Sulphas	1½	1	...	...	...	...	...	...	12	8
40	Cocainæ Hydrochloras	...	...	...	...	...	...	...	...	1½	1
41	Collodium	2	1	...	...	...	...	...	...	12	8

DIX C.

—

are kept at the Government Medical Stores, Rangoon, put up ready for indents.

81, and 83.]

plied by Avoirdupois weight of 437·5 grains, or 16 drachms to the ounce. All other measure.

TRATED HYDROCHLORIC, NITRIC, OR SULPHURIC ACID ISSUED.

Oz.								Drs.						Grains.				
6	4	2	...	...	...	...	...	...	...	...	...	...	...	...	...	...	...	...
6	4	3	2	1	...	...	...	...	...	...	...	...	...	...	...	...	...	...
6	5	4	3	2	1	...	...	...	...	...	...	...	...	...	...	...	...	...
4	3	2	1	...	...	...	...	...	...	...	...	...	...	...	...	...	...	...
6	4	3	2	...	...	...	...	...	...	...	...	...	...	...	...	...	...	...
6	5	4	3	2	...	...	...	...	...	...	...	...	...	...	...	...	...	...
6	5	4	3	2	1	...	...	...	...	...	...	...	...	...	...	...	...	...
6	5	4	3	2	1	...	...	8	4	...	...	...	...	...	...	...	...	...
6	5	4	3	2	1	...	...	8	4	...	...	...	...	...	...	...	...	...
...	...	...	...	...	...	...	...	...	...	...	...	...	...	...	...	...	...	...
3	2	1	...	...	...	...	...	8	4	2	...	...	...	...	...	...	...	...
6	5	4	3	2	1	...	...	...	...	...	...	...	...	...	...	...	...	...
...	...	...	...	...	...	...	...	...	...	...	...	...	...	...	...	...	...	...
6	5	4	3	2	1	...	...	8	4	...	...	...	...	...	...	...	...	...
6	5	4	3	2	1	...	...	...	...	...	...	...	...	...	...	...	...	...
6	5	4	3	2	1	...	...	...	...	...	...	...	...	...	...	...	...	...
6	5	4	3	2	1	...	...	...	...	...	...	...	...	...	...	...	...	...
6	5	4	3	2	1	...	...	...	...	...	...	...	...	...	...	...	...	...
3	2	1	...	...	...	...	...	8	6	4	2	1	...	...	...	...	...	...
6	5	4	...	...	...	...	...	...	...	...	...	...	...	...	...	...	...	...
...	...	...	...	...	...	...	...	8	4	2	1	...	10	5	...	...	...	...
...	...	...	...	...	...	...	...	...	...	...	...	...	...	...	...	...	...	...
8	6	5	4	3	2	1	...	8	6	4	2	...	...	...	...	...	...	...
8	6	5	4	3	2	1	...	8	6	4	2	...	...	...	...	...	...	...
6	4	3	2	1	...	...	...	...	...	...	...	...	...	...	...	...	...	...
...	...	...	...	...	...	...	...	4	3	2	1	...	...	...	20	15	10	5
8	6	5	4	3	2	1	...	...	...	...	...	...	...	...	...	...	...	...
6	5	4	3	2	1	...	...	...	...	...	...	...	...	...	...	...	...	...
6	5	4	3	2	1	...	...	...	...	...	...	...	...	...	...	...	...	...
6	5	4	3	2	1	...	...	...	...	...	...	...	...	...	...	...	...	...
6	4	2	1	...	...	...	...	...	...	...	...	...	...	...	...	...	...	...
2	1½	1	...	...	...	...	...	8	4	2	1	...	...	...	...	...	...	...
6	4	...	...	...	...	...	...	...	...	...	...	...	...	...	...	...	...	...
8	6	5	4	3	2	1	...	8	4	...	...	...	...	...	...	...	...	...
8	6	5	4	3	2	1	...	...	...	...	...	...	...	...	...	...	...	...
6	5	4	3	2	1	...	...	8	4	...	...	...	...	...	...	...	...	...
6	4	2	1	...	...	...	...	...	...	...	...	...	...	...	...	...	...	...
6	4	2	1	...	...	...	...	...	...	...	...	...	...	...	...	...	...	...
½	¼	...	...	...	...	...	...	7	6	5	3	2	1	...	20	15	10	5
6	5	4	3	2	1	...	...	8	4	...	...	...	...	...	...	...	...	...

Table showing the authorized drugs and quantities in which such drugs issue on indents

Serial No.	—	Lbs.									
	I.—Medicines—continued.										
42	Copaiba	4	3	1½	1	...	...	...	...	12	8
43	Creasotum	1	...	...	...	...	...	...	...	12	8
44	Creta Præparata	8	6	5	4	3	2	1½	1	12	8
45	Cubeba	2	1	...	...	..	...	...	...	12	8
46	Cupri Sulphas	2	1½	1	...	..	...	...	...	12	8
47	Emplastrum Belladonnæ ...	2	1	...	...	...	...	...	...	12	8
48	Emplastrum Cantharidis ...	2	1	...	...	...	...	...	...	12	8
49	Emplastrum Hydrarg ...	2	1	...	...	...	...	...	...	12	8
50	Emplastrum Opii	...	...	...	...	...	...	...	...	8	6
51	Emplastrum Picis	1	...	...	...	...	...	...	...	12	8
52	Emplastrum Resinæ ...	...	...	...	...	...	...	...	...	...	...
53	Ergota	...	...	...	...	...	...	...	...	8	6
54	Ergotinum, Bonjeans' ...	...	...	...	...	...	...	...	...	4	3
55	Extractum Belladonnæ ...	2	1	...	...	...	...	...	...	12	8
56	Extractum Cannabis Indica ...	1	...	...	...	...	...	...	...	12	8
57	Extractum Colocynthidis Compositum.	2	1	...	...	...	...	...	...	12	8
58	Extractum Ergotæ Liquidum ...	2	1½	1	...	..	...	...	...	12	8
59	Extractum Filicis Liquidum ...	1	...	...	...	...	...	...	...	12	8
60	Extractum Hyoscyami ...	1	...	...	...	...	...	...	...	12	8
61	Extractum Nucis Vomicæ ...	...	...	...	...	...	...	...	...	4	3
62	Ferri et Ammonii Citras ...	1	...	...	...	...	.	...	...	12	8
63	Ferri et Quininæ Citras ...	1	...	...	...	...	...	...	...	12	8
64	Ferri Sulphas ...	2	1½	1	...	...	...	...	...	12	8
65	Glycerinum	2½	2	1½	1	...	...	...	...	12	8
66	Hydrargyri Iodidum Rubrum ...	1	...	...	...	...	...	...	...	12	8
67	Hydrargyri Oxidum Flavum ...	...	...	...	...	...	...	...	...	2	1
68	Hydrargyri Perchloridum ...	1	...	...	...	...	...	...	...	12	8
69	Hydrargyri Subchloridum ...	4	3	2	1½	1	...	...	...	12	8
70	Hydrargyrum cum Cretæ Saccharata.	1	...	...	...	...	...	...	...	12	8
71	Iodoformum	1½	1	...	...	...	...	...	...	12	8
72	Iodum	1	...	...	...	...	...	...	...	12	8
73	Ipecacuanhæ (Pulvis) ...	4	3	2	1½	1	...	...	...	12	8
74	Ipecac Sine Emetine ...	1	...	...	...	...	...	...	...	12	8
75	Jalapæ (Pulvis)	2	1	...	...	...	...	...	...	12	8
76	Kamala	...	...	...	...	...	...	...	...	8	6
77	Linimentum Belladonnæ ...	4	3	2	1½	1	...	...	...	12	8
78	Linimentum Camphoræ Compositum.	4	3	2	1½	1	...	...	...	12	8
79	Linimentum Saponis ...	4	3	2	1½	1	...	...	...	12	8
80	Liquor Ammoniæ	4	3	2	1½	1	...	...	...	12	8
81	Liquor Arsenicalis	4	3	2	1½	1	...	...	...	12	8
82	Liquor Arsenii et Hydrarg Iodidi	2	1	...	...	...	...	...	...	12	8
83	Liquor Epispasticus ...	1	...	..	...	...	...	...	...	12	8
84	Liquor Ferri Perchloridi Fortior	2	1½	1	...	...	...	...	...	12	8
85	Liquor Opii Sedativus ...	1½	1	...	...	...	...	...	...	12	8
86	Liquor Plumbi Subacetatis ...	2	1½	1	...	...	...	...	...	12	8
87	Liquor Potassæ	4	3	2	1½	1	...	...	...	12	8
88	Liquor Strychnia Hydrochloratis	...	...	...	...	...	...	...	...	8	6
89	Liquor Zinci Chloridi ...	2	1½	1	...	...	...	...	...	12	8
90	Magnesia Levis ...	...	...	...	...	...	...	...	...	12	8

are kept at the Government Medical Stores, Rangoon, put up ready for —continued.

Oz.								Drs.						Grains.					
6	5	4	3	2	1	...	...	...	...	...	...	...	...	...	...	...	...	...	...
6	5	4	3	2	1	...	...	8	4	2	...	...	.	...	...	...	...	...	...
6	5	4	3	2	1	...	...	...	...	...	...	...	...	...	...	...	...	...	...
6	5	4	3	2	...	...	...	...	...	...	...	...	...	...	...	...	...	...	...
6	5	4	3	2	1	...	...	...	...	...	...	...	...	...	...	...	...	...	...
6	5	4	3	2	1	...	...	...	...	...	...	...	...	...	...	...	...	...	...
6	5	4	3	2	1	...	...	8	...	...	...	. .	...	...	...	...	...	...	...
4	2	1	...	...	...	...	...	...	...	...	...	...	...	...	...	...	...	...	...
6	4	2	1	...	...	...	...	...	...	...	...	...	...	...	...	...	...	...	...
..	..	...	...	...	...	...	...	...	...	...	...	...	...	...	...	...	...	...	...
4	2	1	...	...	...	...	...	...	...	...	...	...	...	...	...	...	...	...	...
2	1	...	...	...	...	...	...	8	4	2	1	...	...	...	...	...	...	...	...
6	5	4	3	2	1	...	...	8	4	...	...	...	...	...	...	...	...	...	...
6	5	4	3	2	1	...	...	8	4	2	...	...	...	...	...	...	...	...	...
6	5	4	3	2	1	...	...	8	4	...	...	...	...	...	...	...	...	...	...
6	5	4	3	2	1	...	...	8	4	...	...	...	...	...	...	...	...	...	...
6	5	4	3	2	1	...	...	8	4	...	...	...	...	...	...	...	...	...	...
2	1	...	...	...	...	...	...	8	4	2	...	...	...	...	...	...	...	...	...
6	5	4	3	2	1	...	...	8	4	...	...	...	...	...	...	...	...	...	...
6	5	4	3	2	1	...	...	...	...	...	...	...	...	...	...	...	...	...	...
6	5	4	3	2	1	...	...	8	...	...	...	...	...	...	...	...	...	...	...
6	5	4	3	2	1	...	...	...	...	...	...	...	...	...	...	...	...	...	...
6	5	4	3	2	1	...	...	8	4	3	2	1	...	...	...	...	...	...	...
...	...	...	...	...	...	...	...	8	4	3	2	1	...	...	...	...	...	...	...
6	5	4	3	2	1	...	...	8	4	3	2	1	...	...	...	...	...	...	...
6	5	4	3	2	1	...	...	8	4	3	2	1	...	...	...	...	...	...	...
6	5	4	3	2	1	...	...	8	4	3	2	...	...	...	...	...	...	...	...
6	5	4	3	2	1	...	...	...	...	...	...	...	...	...	...	...	...	...	...
6	5	4	3	2	1	...	...	8	4	3	2	...	...	...	...	...	...	...	...
6	5	4	3	2	1	...	...	...	...	...	...	...	...	...	...	...	...	...	...
6	5	4	3	2	1	...	...	...	...	...	...	...	...	...	...	...	...	...	...
6	5	4	3	2	1	...	...	...	...	...	...	...	...	...	...	...	...	...	...
4	2	1	...	...	...	...	...	...	...	...	...	...	...	...	...	...	...	...	...
6	5	4	3	2	...	...	...	...	...	...	...	...	...	...	...	...	...	...	...
6	5	4	3	2	1	...	...	...	...	...	...	...	...	...	...	...	...	...	...
6	5	4	3	2	...	...	...	...	...	...	.	...	...	...	...	...	...	...	...
6	5	4	3	2	1	...	...	...	...	...	...	...	...	...	...	...	...	...	...
6	5	4	3	2	1	...	...	...	...	...	...	...	...	...	...	...	...	...	...
6	5	4	3	2	1	...	...	4	...	.	...	...	...	...	...	...	...	...	...
6	5	4	3	2	1	...	...	4	...	...	...	...	...	...	...	...	...	...	...
6	5	4	3	2	1	...	...	...	...	...	...	...	...	...	...	...	...	...	...
6	5	4	3	2	1	...	...	...	...	...	...	...	...	...	...	...	...	...	...
6	5	4	3	2	1	...	...	...	...	...	...	...	...	...	...	...	...	...	...
6	5	4	3	2	1	...	...	...	...	...	...	...	...	...	...	...	...	...	...
5	4	3	2	1	..	...	...	...	...	...	.	...	...	...	...	...	...	...	...
6	5	4	3	2	1	...	...	...	...	...	...	...	...	...	...	...	...	...	...
6	5	4	3	2	1	...	...	8	4	...	...	...	...	...	...	...	...	...	...

Table showing the authorized drugs and quantities in which such drugs issue on indents

I.—Medicines—continued.

Serial No.		Lbs.									
91	Magnesii Sulphas ...	8	6	5	4	3	2	1½	1	12	8
92	Manganesii Oxidum Nigrum ...	...	...	...	...	...	...	...	...	4	3
93	Morphinæ Acetas ...	...	...	...	...	...	...	...	...	1½	1
94	Morphinæ Hydrochloras ...	...	...	...	...	...	...	...	...	2	1½
95	Oleatum Hydrargyri ...	...	...	...	...	...	...	...	...	12	8
96	Oleum Anethi ...	...	...	...	...	...	...	...	...	8	6
97	Oleum Anisi ...	...	...	...	...	...	...	...	...	8	6
98	Oleum Arachis Hypogœa ...	4	3	2	1½	1	...	...	...	12	8
99	Oleum Crotonis ...	...	...	...	...	...	...	...	...	2	1
100	Oleum Eucalypti ...	1	...	...	...	...	...	...	...	12	8
101	Oleum Menthæ Piperitæ ...	2	1½	1	...	...	...	...	...	12	10
102	Oleum Morrhuæ ...	4	3	2	1	...	...	...	...	12	8
103	Oleum Ricini ...	...	...	...	...	...	...	...	...	...	...
104	Oleum Rutæ ...	...	...	...	...	...	...	...	...	...	...
105	Oleum Santali ...	2	1½	1	...	...	...	...	...	12	8
106	Oleum Terebinthinæ ...	4	3	2	1½	1	...	...	...	12	8
107	Opium Cake ...	2	1½	1	...	...	...	...	...	12	8
108	Paraffinum Durum ...	4	3	2	1	...	...	...	...	12	8
109	Paraffinum Molle (Vaseline) ...	8	6	5	4	3	2	1½	1	12	8
110	Pepsine ...	...	...	...	...	...	...	...	...	1	...
111	Phenacetin ...	1½	1	...	...	...	...	...	...	12	8
112	Pilula Aloes et Myrrhæ ...	...	...	...	...	...	...	...	...	12	8
113	Pilula Hydrargyri ...	1	...	...	...	...	...	...	...	12	8
114	Pilula Hydrargyri Subchloridi Composita.	1	...	...	...	...	...	...	...	12	8
115	Pilula Rhei Composita ...	1	...	...	...	...	...	...	...	12	8
116	Plumbi Acetas ...	4	3	2	1½	1	...	...	...	12	8
117	Podophylli Resina ...	...	...	...	...	...	...	...	...	1	½
118	Potassii Acetas ...	2	1	...	...	...	...	...	...	12	8
119	Potassii Bicarbonas ...	8	6	5	4	3	2	1½	1	12	8
120	Potassii Bromidum ...	4	3	2	1½	1	...	...	...	12	8
121	Potassii Chloras ...	8	6	5	4	3	2	1½	1	12	8
122	Potassii Iodidum ...	8	6	5	4	3	2	1½	1	12	8
123	Potassi Nitras ...	8	6	5	4	3	2	1½	1	12	8
124	Potassii Permanganas ...	2	1½	1	...	...	...	...	...	12	8
125	Potassii Tartras Acida ...	4	2	1½	1	...	...	...	...	12	8
126	Pulvis Cretæ Aromaticus ...	2	1½	1	...	...	...	...	...	12	8
127	Pulvis Cretæ Aromaticus cum Opio.	2	1½	1	...	...	...	...	...	12	8
128	Pulvis Ipecacuanhæ Compositus	4	3	2	1½	1	...	...	...	12	8
129	Pulvis Jalapæ Compositus ...	4	3	2	1½	1	...	...	...	12	8
130	Pulvis Kino Compositus ...	...	...	...	...	...	...	...	...	12	8
131	Pulvis Rhei Compositus ...	4	3	2	1½	1	...	...	...	12	8
132	Quassiæ Lignum ...	2	1½	1	...	...	...	...	...	12	8
133	Quininæ Sulphas ...	...	2½	2	1½	1	...	...	...	12	10
134	Rhei Radicis (Pulvis) ...	2	1½	1	...	...	...	...	...	12	8
135	Santoninum ...	1	...	...	...	...	...	...	...	12	8
136	Senega Radix ...	4	3	2	1½	1	...	...	...	12	8
137	Senna Indica ...	4	3	2	1½	1	...	...	...	12	8
138	Soda Tartarata ...	4	3	2	1½	1	...	...	...	12	8
139	Sodii Bicarbonas ...	8	6	5	4	3	2	1½	1	12	8
140	Sodii Salicylas ...	4	3	2	1½	1	...	...	...	12	8

are kept at the Government Medical Stores, Rangoon, put up ready for
—continued.

Oz.								Drs.						Grains.				
6	5	4	3	2	1	...	...	...	...	...	...	...	...	...	...	...	...	...
2	1	...	...	...	...	...	...	...	...	...	...	...	...	...	...	...	...	...
½	¼	...	...	...	...	...	...	12	10	6	3	2	1	...	20	15	10	5 ...
1	½	¼	...	...	...	...	...	12	10	6	3	2	1	...	20	15	10	5 ...
6	5	4	3	2	1	...	...	8	4	...	...	...	...	...	...	...	...	...
5	4	3	2	1	...	...	...	8	4	3	2	1	...	...	...	...	...	...
5	4	3	2	1	...	...	...	8	4	3	2	1	...	...	...	...	...	...
6	4	...	...	...	...	...	...	...	...	...	...	...	...	...	...	...	...	...
...	...	...	...	...	...	...	...	...	4	3	2	1	...	...	...	...	...	...
6	5	4	3	2	1	...	...	...	...	...	...	...	...	...	...	...	...	...
8	6	5	4	3	2	1	...	8	4	3	2	1	...	...	...	...	...	...
6	5	4	3	2	...	...	...	...	...	...	...	...	...	...	...	...	...	...
...	...	...	In bottles of 1 qt. and tins of 1 doz. qts.									...	...	...	...	...	...
...	...	...	...	...	...	...	...	...	...	...	...	...	...	...	...	...	...	...
6	5	4	3	2	1	...	...	8	4	...	...	...	...	...	...	...	...	...
6	5	4	3	2	1	...	...	...	...	...	...	...	...	...	...	...	...	...
6	5	4	3	2	1	...	...	8	4	2	...	...	...	...	...	...	...	...
6	4	2	1	...	...	...	...	...	...	...	...	...	...	...	...	...	...	...
6	5	4	3	2	1	...	...	...	...	...	...	...	...	...	...	...	...	...
...	...	...	...	...	...	...	...	8	4	2	...	...	...	...	...	...	...	...
6	5	4	3	2	1	...	...	...	...	...	...	...	...	...	...	...	...	...
6	5	4	3	2	1	...	...	8	4	...	...	...	...	...	...	...	...	...
6	5	4	3	2	1	...	...	8	4	...	...	...	...	...	...	...	...	...
6	5	4	3	2	1	...	...	8	4	...	...	...	...	...	...	...	...	...
6	5	4	3	2	1	...	...	8	4	...	...	...	...	...	...	...	...	...
6	5	4	3	2	1	...	...	8	4	...	...	...	...	...	...	...	...	...
...	...	...	...	...	...	...	...	9	6	5	3	2	1	...	...	...	...	...
6	5	4	3	2	1	...	...	8	...	...	...	...	...	...	...	...	...	...
6	5	4	3	2	1	...	...	...	...	...	...	...	...	...	...	...	...	...
6	5	4	3	2	1	...	...	...	...	...	...	...	...	...	...	...	...	...
6	5	4	3	2	1	...	...	...	...	...	...	...	...	...	...	...	...	...
6	5	4	3	2	1	...	...	...	...	...	...	...	...	...	...	...	...	...
6	5	4	3	2	1	...	...	...	...	...	...	...	...	...	...	...	...	...
6	5	4	3	2	1	...	...	...	...	...	...	...	...	...	...	...	...	...
6	5	4	3	2	1	...	...	...	...	...	...	...	...	...	...	...	...	...
6	5	4	3	2	1	...	...	...	...	...	...	...	...	...	...	...	...	...
6	5	4	3	2	1	...	...	...	...	...	...	...	...	...	...	...	...	...
6	5	4	3	2	1	...	...	...	...	...	...	...	...	...	...	...	...	...
6	5	4	3	2	...	...	...	...	...	...	...	...	...	...	...	...	...	...
8	6	4	3	2	...	...	...	...	...	...	...	...	...	...	...	...	...	...
6	5	4	2	1	1	...	...	...	...	...	...	...	...	...	...	...	...	...
6	5	4	3	2	1	...	...	8	4	3	2	1	...	...	...	...	...	...
6	5	4	3	2	1	...	...	...	...	...	...	...	...	...	...	...	...	...
6	4	...	...	...	...	...	...	...	...	...	...	...	...	...	...	...	...	...
6	5	4	3	2	1	...	...	...	...	...	...	...	...	...	...	...	...	...
6	5	4	3	2	1	...	...	...	...	...	...	...	...	...	...	...	...	...
6	5	4	3	2	1	...	...	...	...	...	...	...	...	...	...	...	...	...

Table showing the authorized drugs and quantities in which such drugs issue on

Serial No.						Lbs.						

I.—Medicines—concluded.

141	Spirits Ætheris	4	3	2	1½	1	...	...	...	12	8
142	Spirits Ætheris Nitrosi ...	4	3	2	1½	1	...	...	...	12	8
143	Spirits Ammoniæ Aromaticæ ..	4	3	2	1½	1	...	...	...	12	8
144	Spirits Rectificatus	4	3	2	1½	1	...	...	...	12	8
145	Strychnina		...	...	...	...	...	...	...	1	...
146	Sulphonal	1	...	...	...	...	...	...	...	12	8
147	Sulphur Sublimatum ...	4	3	2	1½	1	...	...	...	12	8
148	Syrupus Ferri Iodidi ...	4	3	2	1½	1	...	...	...	12	8
149	Tinctura Aconiti ...	1	...	...	...	...	...	...	...	12	8
150	Tinctura Aloes ...	1	...	...	...	...	...	...	...	12	8
151	Tinctura Arnicæ	1	...	...	...	...	...	...	...	14	12
152	Tinctura Asafœtida	4	3	2	1½	1	...	...	...	14	12
153	Tinctura Aurauti	2	1½	1	...	...	...	...	...	14	12
154	Tinctura Belladonnæ ...	1	...	...	...	...	...	...	...	14	12
155	Tinctura Benzoini Composita ...	2	1½	1	...	...	...	...	...	12	10
156	Tinctura Calumbæ ...	4	3	2	1½	1	...	...	...	14	12
157	Tinctura Camphora Composita	4	3	2	1½	1	...	...	...	14	12
158	Tinctura Cannabis Indica ...	1½	1	.	..	...	..	...	...	14	12
159	Tinctura Capsici	2	1	...	...	...	...	...	...	14	12
160	Tinctura Cardamomi Composita	4	3	2	1½	1	...	...	...	14	12
161	Tinctura Catechu	4	3	2	1½	1	...	...	...	14	12
162	Tinctura Cinchonæ Composita ...	4	3	2	1½	1	...	...	...	14	12
163	Tinctura Chirettæ	...	...	...	...	...	...	...	...	...	...
164	Tinctura Digitalis	4	3	2	1½	1	...	...	...	14	12
165	Tinctura Ferri Perchloridi ...	4	3	2	1½	1	...	...	...	14	12
166	Tinctura Gentianæ Composita ...	4	3	2	1½	1	...	...	...	14	12
167	Tinctura Hyoscyami ...	4	3	2	1½	1	...	...	...	14	12
168	Tinctura Iodi	4	3	2	1½	1	...	...	...	14	12
169	Tinctura Ipecacuanha ...	4	3	2	1½	1	...	...	...	14	12
170	Tinctura Kino	4	3	2	1½	1	...	...	...	12	8
171	Tinctura Lobelia Ætherea ...	2	1½	1	...	...	...	...	...	14	12
172	Tinctura Nucis Vomicæ ...	2	1½	1	...	...	...	...	...	14	12
173	Tinctura Opii	4	3	2	1½	1	...	...	...	14	12
174	Tinctura Quassia	4	3	2	1½	1	...	...	...	14	12
175	Tinctura Rhei	2	1	...	...	...	...	...	...	12	10
176	Tinctura Scillæ	4	3	2	1½	1	...	...	...	14	12
177	Tinctura Senega	2	1	...	...	...	...	...	...	12	8
178	Tinctura Senna	1	...	...	...	...	...	...	...	12	10
179	Tinctura Valerianæ Ammoniata	1	...	...	...	...	...	...	...	12	8
180	Tinctura Zingiberis	4	3	2	1½	1	...	...	...	12	8
181	Unguentum Acidi Borici (Lister's)	8	4	3	2	1½	1	...	...	12	8
182	Unguentum Hydrargyri ...	2	1½	1	...	...	...	...	...	12	8
183	Unguentum Hydrargyri Nitratis	1	...	...	...	...	..	...	...	12	8
184	Unguentum Resinæ ...	4	3	2	1½	1	...	...	...	12	8
185	Unguentum Simplex ...	8	6	5	4	3	2	1½	1	12	8
186	Unguentum Sulphuris ...	8	6	5	4	3	2	1½	1	12	8
187	Vinum Antimoniale ...	1½	1	...	...	...	...	...	...	12	8
188	Vinum Colchici	2	1	...	...	...	...	...	...	12	8
189	Zinci Colchoridum ...	...	...	...	...	...	...	...	...	8	6
190	Zinci Oxidum	2	1½	1	...	...	...	...	...	12	8
191	Zinci Sulphas	4	3	2	1½	1	...	...	...	12	8

Oz.								Drs.							Grains.			
6	5	4	3	2	1	...	...	...	...	...	...	...	...	...	...	...	...	...
6	5	4	3	2	1	...	...	...	...	...	...	...	...	...	...	...	...	...
6	5	4	3	2	1	...	...	...	...	...	...	...	...	...	...	...	...	...
6	5	4	3	2	1	...	...	...	...	...	...	...	...	...	...	...	...	...
...	...	...	...	...	...	...	8	6	4	3	2	1	...	20	15	10	5	...
6	5	4	3	2	1	...	...	...	...	...	...	...	...	...	...	...	...	...
6	5	4	3	2	...	...	...	...	...	...	...	...	...	...	...	...	...	...
6	5	4	3	2	1	...	...	4	...	...	...	...	...	...	...	...	...	...
6	5	4	3	2	1	...	...	...	...	...	...	...	...	...	...	...	...	...
10	8	6	5	4	3	2	1	...	...	...	...	...	...	...	...	...	...	...
10	8	6	5	4	3	2	1	...	...	...	...	...	...	...	...	...	...	...
10	8	6	5	4	3	2	1	...	...	...	...	...	...	...	...	...	...	...
10	8	6	5	4	3	2	1	...	...	...	...	...	...	...	...	...	...	...
8	6	5	4	3	2	1	...	...	...	...	...	...	...	...	...	...	...	...
10	8	6	5	4	3	2	1	...	...	...	...	...	...	...	...	...	...	...
10	8	6	5	4	3	2	1	...	...	...	...	...	...	...	...	...	...	...
10	8	6	5	4	3	2	1	...	...	...	...	...	...	...	...	...	...	...
10	8	6	5	4	3	2	1	...	...	...	...	...	...	...	...	...	...	...
10	8	6	5	4	3	2	1	...	...	...	...	...	...	...	...	...	...	...
10	8	6	5	4	3	2	1	...	...	...	...	...	...	...	...	...	...	...
10	8	6	5	4	3	2	1	...	...	...	...	...	...	...	...	...	...	...
...	...	...	...	...	...	...	...	...	...	...	...	...	...	...	...	...	...	...
10	8	6	5	4	3	2	1	...	...	...	...	...	...	...	...	...	...	...
10	8	6	5	4	3	2	1	...	...	...	...	...	...	...	...	...	...	...
10	8	6	5	4	3	2	1	...	...	...	...	...	...	...	...	...	...	...
10	8	6	5	4	3	2	1	...	...	...	...	...	...	...	...	...	...	...
10	8	6	5	4	3	2	1	...	...	...	...	...	...	...	...	...	...	...
10	8	6	5	4	3	2	1	...	...	...	...	...	...	...	...	...	...	...
6*	5	4	3	2	1	...	...	...	...	...	...	...	...	...	...	...	...	...
10	8	6	5	4	3	2	1	...	...	...	...	...	...	...	...	...	...	...
10	8	6	5	4	3	2	1	...	...	...	...	...	...	...	...	...	...	...
10	8	6	5	4	3	2	1	...	...	...	...	...	...	...	...	...	...	...
10	8	6	5	4	3	3	1	...	...	...	...	...	...	...	...	...	...	...
8	6	5	4	3	2	1	...	...	...	...	...	...	...	...	...	...	...	...
10	8	6	5	4	3	2	1	...	...	...	...	...	...	...	...	...	...	...
6	4	2	1	...	...	...	...	...	...	...	...	...	...	...	...	...	...	...
8	6	4	3	2	1	...	...	...	...	...	...	...	...	...	...	...	...	...
6	4	3	2	1	...	...	...	...	...	...	...	...	...	...	...	...	...	...
6	5	4	3	2	1	...	...	...	...	...	...	...	...	...	...	...	...	...
6	5	4	3	2	...	...	...	...	...	...	...	...	...	...	...	...	...	...
6	5	4	3	2	1	...	...	8	...	...	...	...	...	...	...	...	...	...
6	5	4	3	2	1	...	...	8	...	...	...	...	...	...	...	...	...	...
6	5	4	3	2	1	...	...	...	...	...	...	...	...	...	...	...	...	...
6	5	4	..	...	...	...	...	...	...	...	...	...	...	...	...	...	...	...
6	5	4	...	...	...	...	...	...	...	...	...	...	...	...	...	...	...	...
6	5	4	3	2	1	...	...	...	...	...	...	...	...	...	...	...	...	...
6	5	4	3	2	1	...	...	...	...	...	...	...	...	...	...	...	...	...
5	4	3	2	1	...	...	...	...	...	...	...	...	...	...	...	...	...	...
6	5	4	3	2	1	...	...	...	...	...	...	...	...	...	...	...	...	...
6	5	4	3	2	1	...	...	...	...	...	...	...	...	...	...	...	...	...

Minimum scale of equipment for Civil Hospitals and Dispensaries.

[See Rule 97.]

Article.	Civil dispensary.	Civil hospital up to ten beds; articles to be increased proportionately as required.
SECTION I.		
Surgical Instruments.		
Bougies, metallic, sets of 12	...	1
Catheters, elastic, gum, sets of 12	1	1
„ German silver, female	1	1
„ „ „ male, sets of 12	1	1
Instruments, cupping case, large	1	1
Amputating, Regulation, D	1	1
Instruments, dressing, pocket-case, large ...	1	1
„ hydrocele	1	1
„ midwifery	1	1
„ post-mortem case, large *	1	1
„ tooth-case	1	1
Lancets, abscess, Symes'	1	1
„ bleeding, in case	1	1
„ vaccinating	1	1
Needles, common curved	6	6
„ exploring	1	1
Trocars, set, in case	1	1
Tubes, tracheotomy, dilating, sets	1	1
SECTION II.		
Medical and Surgical Appliances.		
Apparatus, electro-magnetic	1	1
„ irrigation, without vessel	1	1
„ urinometer	...	1
Inhaler, Mudge's	...	1
„ spare tubes	...	1
Ligature, flax	1	1
„ silk	1	1
Pencils, camel-hair	2	2
Pumps, stomach, Reid's	1	1
Specula recti	1	1
Splints, common, long, Dessault's, sets ...	2	2
„ „ sets	1	1
Stethoscopes	1	1
Syringes, enema, Higginson's, with vaginal tube ...	1	1
„ glass, urethra, male, ½ oz.	1	1

* See Rule 100.

Minimum scale of equipment for Civil Hospitals and Dispensaries—concluded.

Article.	Civil dispensary.	Civil hospital up to ten beds; articles to be increased proportionately as required.
SECTION II—concluded.		
Medical and Surgical Appliances—concluded.		
Syringes, hypodermic	1	1
Thermometers, clinical, self-registering	1	1
Torniquets, Esmarch's, complete	1	1
SECTION III.		
Sundries.		
Basins, pus	2	2
„ dressing, metal, sets of five	1	1
Corkscrews	1	1
Funnels, composition, of sizes	1	1
Measure glass, 4 oz.	1	1
„ „ 2 oz.	1	1
„ „ minim	1	1
„ pewter, 2 oz.	1	1
„ double, ½ oz. and 1 oz.	1	1
Pans, bed, earthenware, shape	...	1
Pestles and mortars, brass	1	1
„ „ glass	1	1
„ „ wedgewood	1	1
Pots, decoction tin	1	1
„ infusion tin	1	1
Rods, glass	1	1
Scales and weights, from 2 lb. downwards ...	1	1
„ „ grains and drachm pillar ...	1	1
Scissors, sharp	1	1
Slabs, wedgewood, of sizes	1	1
Spatula, bolus	2	2
„ spreading	1	1
Urinals, glass	...	1
Weights, grain, spare, sets	1	1

APPENDIX E.

List of Poisons.

[See Rules 92 to 96.]

Acetum cantharidis.
Acidum arseniosum.
„   carbolicum.
„   „   liquefactum.
„   chromicum.
„   chrysophanicum.
„   hydrochloricum.
„   hydrocianicum dilutum.
„   nitricum.
„   phosphoricum concentratum.
„   sulphuricum.
Aconiti radix.
Aconitina.
Æther purus.
Alcohol amylicum.
Aloin.
Amyl nitris.
Antimonii oxidum.
Antimonium sulphuratum.
„   tartaratum.
Antiseptic solution.
Apomorphinæ hydrochloras.
Argenti et potassi nitras.
Argenti nitras.
Argenti oxidum.
Arsenii iodidum.
Atropina.
„   sulphas.
Belladonnæ radix.
Bromum.
Butyl chloral hydras.
Cantharis.
Cerii oxalas.
Chloral hydras.
Chlorodyne.
Chloroformum.
Chrysarobinum.
Cocainæ hydrochloras.
Codeina.
Collodium vesicans.
Creasotum.
Cupri nitras.
Cupri sulphas.
Digitalis folia.
Ecballii fructus.
Elaterinum.
Elaterium.
Emplastrum cantharidis.
Ergota.
Ergotinum.
Extractum aconiti.
„   belladonnæ.
„   alcoholium.

Extractum cannabis indica.
„   colchici.
„   „   aceticum.
„   ergotæ liquidum.
„   gelsemii alcoholicum.
„   nucis vomicæ.
„   opii.
„   „   liquidum.
„   physostigmatis.
„   stramonii.
Ferri arsenias.
Hydrargyri iodidum rubrum.
„   oxidum flavum.
„   „   rubrum.
„   perchloridum.
„   persulphas.
Hydrargyrum ammoniatum.
Injectio apomorphinæ hypodermica.
„   ergotini hypodermica.
„   morphinæ hypodermica.
Iodoformum.
Iodum.
Jalapine.
Lamellæ atropinæ.
„   cocainæ.
„   physostigminæ.
Linimentum aconiti.
„   belladonnæ.
„   crotonis.
„   opii.
Liquor acidi chromici.
„   ammoniæ fortior.
„   arsenicalis.
„   arsenic hydrochloricus.
„   atropinæ sulphatis.
„   epispasticus.
„   hydrargyri nitratis acidus.
„   morphinæ acetatis.
„   „   bimeconatis.
„   „   hydrochloratis.
„   opii sedativus.
„   sodii arseniatis.
„   strychninæ hydrochloratis.
„   zinci chloridi.
Menthol.
Morphinæ acetas.
„   hydrochloras.
„   sulphas.
Nux vomica.
Oleum crotonis.
„   rutæ.
„   sabinæ.
Opium.
Phosphorus.

Physostigmatis semen.
Physostigmina.
Pilocarpinæ nitras.
Plumbi acetas.
Podophylli resina.
Potassa caustica.
Potassi cyanidum.
„ ferrocyanidum.
Pulvis opii compositus.
Pyroxylin.
Sabinæ cacumina.
Soda caustica.
Sodi arsenias.
Sodii valerianas.
Stramonii semina.
Strychnina.
Tabellæ nitroglyceriui.
Thymol.
Tinctura aconiti.

Tinctura belladonnæ.
„ chloroformi et morphinæ.
„ cannabis indica.
„ digitalis.
„ ergotæ.
„ gelsemii.
„ iodi.
„ lobeliæ ætherea.
„ opii.
„ „ ammoniata.
„ nucis vomicæ.
„ sabinæ.
Unguentum aconitinæ.
„ atropinæ.
„ belladonnæ.
„ sabinæ.
Veratri viridis rhizoma.
Veratrina.
Vinum opii.
Zinci chloridum.

APPENDIX F.

[See Rule 190.]

Local Government (Judicial Department) Circular No. 98 of 1896, dated the 12th October 1896.

THE appended extract from Home Department Resolution No. 16 (Medical)-576—86, dated the 10th July 1896, is circulated for the information and guidance of all Magistrates, Judges of Civil Courts, Medical Officers, Officers of the Jail Department, and the Superintendent and Committee of Visitors, Rangoon Lunatic Asylum. In accordance with the orders of the Government of India, the Chief Commissioner directs that the form of Medical History sheet appended hereto* shall be used in all cases in which—

Medical History sheet for lunatics. Criminal lunatics not to be sent to asylum without orders of local Government.

(i) a Magistrate (sections 4 and 5 of Act XXXVI of 1858) or the Judge of a Civil Court (section 8 of Act XXXVI of 1858) orders a lunatic to be received into an asylum ;

(ii) a Magistrate or Court applies for the orders of the local Government under section 466 or section 471 of the Code of Civil Procedure, 1882 ; or

(iii) the orders of the local Government are applied for under section 31 (*I*) of the Prisoners Act, 1871, as amended by Act X of 1886.

2. Civil Surgeons and other Medical Officers are directed to furnish Magistrates and Courts with all the assistance in their power in all cases in which the Medical History sheet has to be filled in.

3. The Medical History sheet now prescribed will be substituted for Form Medl.—17 L.A. in the Medical Guard-book of forms, and copies of it may be obtained on indent in the usual manner.

---

* *See* pages cii and ciii, Appendix I.

4. Circular No. 73 of 1895* is hereby cancelled, and the attention of Magistrates is invited to the orders contained in paragraph 4 of the appended Resolution prescribing the scope of the further inquiries which Magistrates may be required to make at the instance of the Superintendent of the Lunatic Asylum and the manner in which such inquiries should be conducted.

5. Criminal lunatics should not be sent to the asylum without the orders of the Chief Commissioner under section 466 or section 471, Code of Criminal Procedure. Pending the Chief Commissioner's orders the lunatics should be remanded to jail.†

Extract from the Proceedings of the Government of India in the Home Department,—No. 16 (Medical)-576—86, under date Simla, the 10th July 1896.

*   *   *   *

2. * * * The object which the Governor-General in Council had in view was that in every case in which a patient is received into a lunatic asylum a complete medical history of his case should be received with him. Though this end is attained in those cases in which the statute law requires a medical history sheet, there is in the majority of cases no statutory obligation to furnish the Superintendent of the asylum with the previous history of the patient.

The different circumstances under which lunatics may be received and confined in an asylum are the following :—

(a) Under section 4 of Act XXXVI of 1858 the Magistrate has power to send to an asylum lunatics found wandering at large, who are believed to be dangerous by reason of their lunacy. No person can be sent to a lunatic asylum under the provisions of this section without a certificate from a Medical Officer in Form A of the schedule attached to the Act.

(b) Under section 5 of the Act the Magistrate can also order a lunatic, who is not under proper care or control, or who is cruelly treated or neglected, to be received into an asylum. In this case also a certificate in Form A is required. The Act does not prescribe the form of the Magistrate's order, but it is customary for a descriptive-roll to be sent with the certificate in Form A.

(c) Under section 466 of the Criminal Procedure Code, when an accused person is found to be of unsound mind and incapable of making his defence, and the case is one in which bail may not be taken, or if sufficient security is not given, the local Government may, on the report of the Magistrate or Court, order the accused to be confined in a lunatic asylum, and the Magistrate or Court has to give effect to such order.

(d) When the judgment of the Magistrate or Court states that an accused person committed the act alleged against

---

* Magisterial inquiries into causes of insanity.

† For rules concerning the detention of supposed civil lunatics under observation, *see* page 9 of Burma Rules Manual (Edition 1897).

him, but acquits him upon the ground that at the time when he committed it he was, by reason of unsoundness of mind, incapable of knowing the nature of the act alleged as constituting the offence, or that it was wrong or contrary to law, the local Government may (section 471, Criminal Procedure Code) order such person to be confined in a lunatic asylum.

(e) Under section 31 (*1*) of the Prisoners Act, V of 1871, as amended by Act X of 1886, whenever it appears to a local Government that any person detained or imprisoned under any order or sentence of a Magistrate or Court is of unsound mind, it may, by a warrant setting forth the grounds of belief that the person is of unsound mind, order his removal to a lunatic asylum within the territories subject to it.

(f) In all other cases outside the Presidency towns (section 8 of Act XXXVI of 1858) an order of the Civil Court is required, which has to be accompanied by a certificate in Form A.

(g) In the Presidency towns (section 7 of Act XXXVI of 1858) a lunatic may be received into an asylum with an order under Form B of the schedule appended to the Act, if he has been found to be a lunatic by inquisition or under an inquiry directed by order of one of the Courts of Judicature established by Royal Charter, and, if not so found, with an order in Form B, supported by a certificate in Form A, signed by two physicians or surgeons, one of whom must be a Presidency Surgeon or a Surgeon in the employ of Government.

3. The cases, therefore, in which it is necessary, by executive order, to direct that a medical history sheet shall be forwarded with the lunatic to the asylum are when he is sent there—

   (i) under section 4 or section 5 of Act XXXVI of 1858,
   (ii) under section 466 or section 471 of the Code of Criminal Procedure,
   (iii) under section 31 (*1*) of the Prisoners Act, V of 1871, as amended by Act X of 1886, or
   (iv) under section 8 of Act XXXVI of 1858 by order of a Civil Court.

The only cases in which the statute law requires any statement of particulars regarding a lunatic to be forwarded with him to the asylum are those in which patients are sent to a lunatic asylum under section 7 of the Act, and the result is that in the great majority of cases there is nothing to ensure that the Superintendent of a Lunatic Asylum shall receive any detailed information about a patient whom he is required to admit into an asylum. The Governor-General in Council is of opinion that a medical statement should in every instance accompany a patient. It appears, however, to His Excellency in Council that this object may be secured without legislation, and he has accordingly had the form of medical history sheet which is appended to this Resolution prepared and desires that it be adopted in all cases in which local Governments and Administrations order [sections 466 and 471 of the Criminal Procedure Code or under section 31 (1) of Act V of 1871 as amended by Act X of 1886] that an accused or convicted person should be confined in a lunatic asylum. The Governor-General in Council will also be glad if, in order to secure uniformity, instructions may, with the concurrence of the High Courts, be issued to secure that the same form is used in all cases in which a Magistrate (sections 4 and 5 of Act XXXVI of 1858) or the Judge of a Civil Court (section 8 of Act XXXVI of 1858) may order a lunatic to be received into an asylum. The Calcutta High Court will be separately moved to issue any instructions which may be considered necessary to secure the object in view in Bengal.

4. Heading 18 of the form attached to this Resolution corresponds to the heading "Supposed cause" in the statement attached to Form B in the schedule to Act XXXVI of 1858. As pointed out in paragraph 4 of the circular dated 31st July last, if the person filling in either of the two forms, one or other of which will in future accompany a lunatic, is in doubt as to the cause of insanity, and is unable to form an opinion, he can abstain from recording the "supposed" cause, or, if he holds an opinion as to the cause but is doubtful of its correctness, he may note the facts on which his opinion is based. Should the Superintendent find, after observation of the lunatic, that the opinion given is probably incorrect, it can always be corrected in the entries in the case-book. It was in order to meet the case of lunatics regarding whom the Superintendent has reason to believe that an incorrect entry has been made and desires to have his own opinion tested by such facts as further inquiry may elicit that the Governor-General in Council proposed that the Superintendent should be authorized to ask the Magistrate to make further inquiry into the cause of insanity or upon any other point regarding which the information given in the statement previously furnished was obscure or obviously incorrect, and that the Magistrate should thereupon be required to take such further evidence as might be available, and to forward the record of his proceedings to the Superintendent of the asylum. The suggestion has met with general approval, but in regard to the rule proposed by the Government of India upon this point—namely, that this further inquiry should be made by the Magistrate himself or by a Subordinate Magistrate and not entrusted to the police or other subordinate officials—the Government of Bengal has expressed a doubt whether it can properly be prescribed as an invariable rule, as the facts of the personal history of a lunatic, if discoverable at all, can best be found out by the police or other subordinates. The Government of India agree with this criticism. In their opinion it should always be open to a Magistrate to employ the agency of the police or other subordinate officials in particular details · connected with his inquiry, but in every case the Magistrate should be personally responsible for the proper conduct of the inquiry.

* * * *

APPENDIX G.

Directions for the maintenance of Registers, Forms, etc., and the correct preparation of Returns.

I.—REGISTER OF OUT-PATIENTS.

1. *Column 2.*—The yearly number commences with the first *new* case attending for treatment and proceeds regularly through the twelve months of the calendar year. The last number registered under this column up to midnight of the 31st December will represent the number of *admissions* during the year.

2. *Column 3.*—The daily number commences and ends with each day. It is only to be prefixed to the name or yearly number of each case applying either personally or through a friend for treatment or to report result. The number attending on one day is the *total attendance* for that day ; such as have received medicines for that day on a previous date, but have not attended personally or through a friend, should not be included in calculating the total daily attendance.

3. *Column 4.*—New cases only should be entered by name ; all others by the yearly number and date of first visit.

4. *Column 6.*—The various classes should be represented by the following initial letters :—

H. for Hindu ; M. for Mussulman ; B., Burmese ; E., European ; Eu., Eurasian ; O. C., other classes.

5. *Column 7.*—The sex should be indicated by the initial letters M. for Male, F., Female ; and C., Children.

NOTE.—*All under 10 years of age should be classed as children.*

6. *Column 8.*—In this column the occupation or trade followed by the patient should be entered.

7. *Column 9.*—Under this column give the permanent residence of the patient, and if he has been taken ill in a place other than his permanent residence, the fact should be noted in the column of remarks.

8. *Columns 10, 11, 12 and 13* are intended to facilitate reference for the monthly returns and to check the numbers registered. The sum of the totals under these columns should equal the number registered under column 3.

9. *Column 14.*—Diseases, injuries, operations, &c., are to be denominated strictly in accordance with the name and number specified in the *Nomenclature of Diseases*, and no name should be used that does not appear in that book. If a variety or subdivision is to be recorded, the letter or figure must follow the number of the main disease.

10. *Column 15.*—The treatment prescribed should be legibly and concisely entered.

11. *Column 16.*—The result, when known, should be entered here. This column may also be used for other purposes of reference, such as entering the number and date under which a case is received from the police, &c.

II.—Ticket for Out-door Patients.

1. This ticket is to be given to each new case attending, with a request that it may be produced at every subsequent visit to the dispensary.

2. If a ticket is not produced at a subsequent visit, the patient should be supplied with a copy of his ticket and should be attended to, but he should be warned to be careful not to forget or lose it.

III.—Register of In-patients.

1. *Column 2.*—The yearly number commences with the first *new* case admitted on the 1st January. The number registered against the last new patient admitted before midnight on the 31st December will indicate the number of cases admitted during the calendar year.

2. *Column 3.*—All cases remaining after midnight on the 31st December are to be entered on the 1st January of the following year by citation of the previous year's yearly number and date.

3. *Columns 5 and 6.*—The classification and the abbreviations used should be the same as prescribed for columns 6 and 7 of the Register of Out-patients, *see* page xxiii.

4. *Columns 7, 8 and 9.*—*See* instructions for columns 8, 9 and 14 of the Register of Out-patients on page xxiii.

5. *Columns 10 to 13.*—Special care should be taken to enter the "result" accurately. No case in which a cure has not been completely effected from a medical point of view should be entered as "cured" under column 10. Cases in which marked amelioration in present symptoms has taken place should be shown as "relieved" under column 11. All cases not returned as "cured," "relieved," or "died" should be shown as "Discharged otherwise" under column 13 and a reference letter or sign should be used to indicate the mode of discharge, thus:

U or (1) would mean "unrelieved."
A or (2) would mean "absconded."
R or (3) would mean "removed by friends," &c.

6. *Column 14.*—To be filled in only after the close of the month.

7. *Column 15.*—Particulars as to cause of complaint, the condition of the patient at the time of discharge, especially in the case of "Discharged otherwise," or other note of interest may be made in this column.

IV.—General Instructions regarding Registers of Out- and In-patients.

1. Lines should be ruled as shown in the forms given in Appendix I in the Register of Out-patients after the close of each day and in the Register of In-patients after the close of each month, and the total should be carefully recorded under each column. This

should be always done by the Medical Officer in charge, and should not be entrusted to a clerk, compounder, or other subordinate.

2. When police cases are admitted, they should be distinguished by the letter P being placed after the yearly number as shown in the forms in Appendix I.

3. For the purposes of the monthly returns all out-patients who do not return within the first seven days of the month following that for which the return is being prepared should be considered as discharged.

4. When a patient admitted for any disease is attacked by another unconnected with the former and before complete recovery from it, a fresh admission will be recorded for the second disease, the case being shown as "Discharged otherwise" against the previous entry, a brief reference being made against each entry in the column for remarks.

5. When a patient is admitted suffering from two diseases at the same time, the case should be shown under the more serious one; should the second disease remain after recovery from the first, the patient will be discharged under the first entry and re-admitted for the second disease.

6. In the case of a patient admitted for a disease which further observation or *post-mortem* examination may prove to have been so entered through an error in diagnosis, the entry should be scored out and the correct disease entered in red ink, so that the case may continue under the original number, and only one admission be reckoned on it.

7. No case of cholera should be discharged for re-admission until every symptom, either directly or indirectly due to the disease, has disappeared.

V.—THE BED-HEAD TICKET AND TEMPERATURE CHART.

1. This ticket is to be hung in a frame above the bed of every in-patient and takes the place of a case book. Full details of temperature, symptoms, progress and treatment are to be recorded each day.

2. On the termination of each case, the ticket should be filed, the tickets of one year being kept together as a record, and should be carefully arranged in regular order by date of admission and month.

VI.—THE DIET SHEET.

1. A diet sheet should be placed in each frame holding the bed-head ticket, and the diet ordered each day should be carefully entered.

If any extras are to be issued, the necessary entry showing the amount ordered should be entered in the appropriate column. This sheet should be attended to each morning immediately after the record on the bed-head ticket has been made.

2. At the end of the month the sheet should be withdrawn from the frame and used as a voucher in support of the entries in the Daily Register of Diets, *see* page xxvi.

VII.—The Daily Register of Diets.

1. This register should be written up daily, the number of diets issued being obtained from the diet-sheets of the patients under treatment.

2. All particulars required by the register should be carefully filled in. Should an entry not be necessary in any of the columns of the register a line should be drawn across it to indicate that no issues were made on that date to the class of patients or of diets or articles named in the first column.

VIII.—Table of Diets.

1.—*Scale of Diets for Europeans.*

Items.	Quantity.	
	lb.	oz.
Full Diet.		
Tea	0	1
Milk	0	6
Sugar	0	3
Bread	1	0
Butter	0	1
Beef	0	12
Vegetables, including 8 oz. potatoes	0	12
Salt	0	1
Onions and pepper	0	0½
Half Diet.		
Tea	0	1
Milk	0	6
Sugar	0	3
Bread	0	8
Butter	0	1
Beef	0	8
Vegetables, including 6 oz. potatoes	0	9
Salt	0	0½
Onions and pepper	0	0½
Milk Diet.		
Tea	0	1
Milk	0	24
Sugar	0	4
Bread	0	8
Butter	0	1
Sago	0	1
Spoon Diet.		
Tea	0	1
Milk	0	12
Sugar	0	3
Sago	0	2

2.—*Scale of Diets for Natives.*

Items.					Quantity.	
					lb.	oz.
Full Diet.						
Rice ...	...	...	...	...	0	24
Beef, fish or dal	...	...	...	...	0	6
Oil or ghee ...	...	...	...	...	0	1
Vegetables ...	...	...	...	...	0	8
Salt ...	...	...	...	...	0	0½
Condiments ...	...	...	...	...	0	0¼
Half Diet.						
Rice ...	...	...	...	...	0	12
Beef, fish or dal ...	...	...	...	...	0	4
Oil or ghee ...	...	...	...	...	0	0½
Vegetables ...	...	...	...	...	0	4
Salt ...	...	...	...	...	0	0½
Condiments ...	...	...	...	...	0	0¼
Milk Diet.						
Milk ...	...	...	...	...	0	24
Bread or rice ...	...	...	...	...	0	8
Sago ...	...	...	...	...	0	1
Sugar ...	...	...	...	...	0	3
Spoon Diet.						
Milk ...	...	...	...	...	0	20
Cornflour, sago, or arrowroot		...	...	...	0	4
Sugar ...	...	...	...	...	0	2

Note.—The scale for the several diets given in the above tables should be strict-ly adhered to. Extras may be issued to patients receiving milk and spoon diets at the discretion of the Medical Officer in charge. Extras should not be allowed with full diets.

IX.—Register of Operations.

1. No distinction need be observed to indicate whether the num-ber entered in the second column refers to the Register of in- or of out-patients, as the number itself and the date on which the operation was performed will afford sufficient clues. This column should on no account be left blank, and the number reproduced should be the yearly and not the daily number given in the hospital registers.

2. The nature and class of operation should be entered in accord-ance with the list given on pages 289 to 323 of the Revised Third Edition of the *Nomenclature of Diseases* published in 1896.

3. Primary operations should be entered in black ink; all second-ary operations in red ink.

4. *See* instructions given for the correct preparation of Statement No. IIIA on pages xxxv to xxxvii. The terminology given in these

instructions and in the list of surgical operations in the *Nomenclature of Diseases* should be strictly adhered to. Should necessity arise for any deviation, the circumstances should be reported by letter to the Administrative Medical Officer with a view to orders being obtained.

X.—Daily Abstract Register showing Class, Sex, Disease, &c., of Patients treated.

1. This abstract is intended to facilitate the preparation of the monthly returns and to show at a glance, for the information of visitors, inspecting officers, and committees, the work done each day during a month at a hospital or dispensary. The information for this abstract should be obtained from the various registers of the institution and written up daily.

2. *Column 1.*—Against the entry "Remaining from previous month" the number of each class, sex and disease of patients who continue to be treated from the previous month should be shown in columns 2b, 3a, 5a and 5b.

3. *Column 2a.*—The number of new out-patients attending personally and those represented by friends should be obtained from the entries under columns 11 and 13 of the Register of Out-patients.

4. *Column 2b.*—Particulars for the various sub-columns should be taken from the entries in columns 6 and 7 of the Register of Out-patients.

5. *Column 2c.*—The number remaining under treatment from day to day are to be added to the total new admissions and shown in this column.

6. *Column 3a.*—Particulars should be obtained from columns 5 and 6 of the Register of In-patients.

7. *Column 3b.*—The number remaining each day under treatment should be added to the number of new admissions of each day and given in this column.

8. *Column 3c.*—The numbers shown under the various headings of this column should, when added together, equal the difference between the aggregate of the daily new admissions and the total under column 3b.

9. *Column 3d.*—The number of persons dieted should be obtained from the Daily Register of Diets.

10. *Column 4.*—The number of police cases treated should be ascertained from the entries in the second column of the Registers of In- and Out-patients.*

11. *Column 5.*—The figures under this column should be equal to the sum of the figures given under the last sub-columns of columns 2b and 3a of this register.

12. *Column 6.*—The particulars for the various headings under this column should be obtained from the Register of Operations.

XI.—Annual Indent for Medicines and Surgical Equipment.

1. Army Hospital Form No. 158, which has now been included in the Guard-book of Medical Forms and numbered Medl.—33 Misc., should be used in all hospitals and dispensaries defined as Class I under Rule 1 of this Manual. Unless otherwise instructed, hospitals and

---

* *See* clause 2, section IV on page xxv.

dispensaries obtaining their supplies from other sources than the Government Medical Stores will use Form Medl.—33A. Misc. (*see* Rule 74).

2. The date from which the 12 months' supplies are required should be distinctly entered in the space provided for this purpose in the heading of the form of indent.

3. Only authorized drugs, instruments and appliances (*see* Appendices C and D) should be included in annual indents.

4. The date required to be given in the heading of the third column of the indent should be that on which the last annual indent was prepared.

5. The quantities received during the period between the date on which the last annual indent was prepared, including all supplies on supplementary and emergent indents, should be entered in the fourth column.

6. The quantities shown as "Remaining in store" should be given as actually found in stock at the time the indent is being prepared.

7. Instruments and appliances ordered to be written off or destroyed by the Administrative Medical Officer since the date of the previous annual indent should be shown in the column headed "Expended during year" and the number and date of the orders of the Administrative Medical Officer should be quoted at the foot of the page on which the name of the article appears.

8. The quantity shown as "Now required" should not exceed the difference between the quantity "Expended" and the quantity "Remaining in store." Should it be found necessary to indent for any quantity in excess of this amount, the reasons for the necessity should be clearly stated (*see* Rule 79 of this Manual).

9. The remaining columns of the form should be left blank.

10. The declaration at the end of the indent should be signed by the Medical Officer in charge of the hospital or dispensary.

11. The certificate required by Rule 75 should be in the form given below and should invariably accompany the indents :—

Form of Certificate.

"I HEREBY certify that the cost of the drugs, instruments. &c., included in this indent does not exceed the amount provided for the purchase of European medicines and instruments in the budget estimate for the year...............
" The amount of this allotment is Rs................."

12. Particular care should be taken to indicate clearly and distinctly in the space provided for the purpose on the last page of the form of indent the full address and route by which it is desired the articles indented for should be sent.

XII.—SUPPLEMENTARY AND EMERGENT INDENTS FOR MEDICINES, INSTRUMENTS, &c.

1. Indents on this form should be sent in only in cases of emergency or when any drugs or instruments ordinarily supplied on annual indents are likely to be expended before the receipt of the annual supply.

2. Form Medl.—33B. Misc. given in Appendix I shall be used by all hospitals and dispensaries.

3. The quantities in stock of articles indented for should be entered as actually found in store at the time of the submission of the indent, and the date on which these quantities were ascertained should be furnished in the space provided in the heading of column 3.

4. Columns 4, 5 and 6 should be filled in as indicated for annual indents, *see* page xxix.

5. The entries in column 7, "Now required," should represent the quantities required to meet demands till the receipt of the annual supply.

6. Instructions given in section XI, pages xxviii and xxix, for the preparation of annual indents should also be observed.

XIII.—Statement showing Losses, Breakages, and Unserviceable Articles.

1. This statement should be submitted in duplicate immediately after the close of the month in which any article of hospital or dispensary equipment is lost, broken or otherwise rendered unserviceable.

2. The name and rank of the person responsible in any way for the loss or damage should invariably be given in the appropriate columns of this statement.

3. The manner in which an article was broken or lost should be clearly and briefly stated in the columns provided for the purpose. The circumstances under which an article has become unserviceable should also be given in this column.

4. The remarks and recommendations of the Civil Surgeon should be entered in that officer's own writing and he should be careful to state whether, in his opinion, the breakage or loss was due to carelessness, accident, or other cause, and make definite recommendations as to whether the article is to be written off or whether it should be replaced at the expense of the party at fault. In the case of unserviceable articles the circumstances under which they have become so should also be explained.

5. The date on which an article lost, broken or rendered unserviceable was brought into use should invariably be given and its value stated approximately. Its condition at the time of its breakage, loss or being damaged should also be furnished.

XIV.—Receipt and Delivery Vouchers.

1. These statements are to be prepared in *triplicate* whenever any article is sent to the Government Medical Stores. Instructions as to the manner in which the vouchers are to be drawn up, the nature of the information to be entered, and the mode of forwarding them are explained in foot-notes on the form and in Rules 111, 112, and 113 of the Manual.

XV.—Surgical Equipment Ledger.

1. This ledger is to be prepared in *duplicate* on the 1st April of each year, and one copy should be forwarded to the Administrative

Medical Officer through the Civil Surgeon of the district; the other copy should be retained on record at the hospital or dispensary from which it is submitted.

2. The number of instruments and appliances, &c., in stock on the 1st April represents the opening balance and should be given under appropriate headings against the entry "Remained 1st April" under "Receipts." Any articles received during the year should be shown under the appropriate headings, and the number and date of the voucher on which they were received should be given in column 1 and the source from which received in column 2.

3. All instruments and appliances returned to the Government Medical Stores for repairs, &c., and all written off as useless, sold, or destroyed, should be shown in the lower half of the form. The number and date of the voucher or order supporting the issue should be shown in column 1, and particulars of issue, as to whom, &c., should be entered in column 2.

4. The difference between the "Total receipts" and "Total issues" given on the last line on each page as "Remaining at date" should represent the closing balance of the year on the 31st March.

5. All instruments and appliances should be correctly described and entered in the ledger in the order in which they are given in Appendix D. Instruments, &c., not given in Appendix D should be named in blank columns at the end of each section.

6. All vouchers in support of receipts and issues should be numbered and entered according to date in the "List of vouchers."

7. At the annual inspection the Civil Surgeon shall enter the results of his examination of the equipment in the form provided on the last page of the ledger headed "Verification of equipment." A copy of this statement shall be forwarded with the copy of the Civil Surgeon's inspection note to the Administrative Medical Officer.

XVI.—$\frac{\text{Monthly}}{\text{Annual}}$ Return of In- and Out-patients.

1. The notes and instructions on this form should always be carefully attended to and all figures should be returned as directed therein.

Out-patients.

2. The *number remaining* should be obtained from the entries under column 4 of the Register of Out-patients or from the daily abstract of patients treated. Only such out-patients as were admitted in the previous year or month, as the case may be, and who returned for treatment within the period prescribed by clause 3 of the General instructions for the correct maintenance of Registers of In- and Out-patients (*see* page xxv) should be considered as remaining from the previous year or month.

3. In preparing the annual return on this form the number remaining from month to month should not be taken into calculation. The number remaining from the year previous to that to which

the return refers should be given in this column, *e.g.*, the number shown as remaining from December in the return for January of the year to which the return relates.

4. The *number admitted* during the year or month should correspond with the sum of the totals entered daily under column 2 of the Register of Out-patients, and the details in the monthly return should correspond with the figures given as the " Total admitted " under column 2*b* of the daily abstract.

In-patients.

5. The *number remaining* should be obtained from the monthly entry under column 14 of the Register of In-patients; thus for the monthly return the figure should be the total given as remaining under this column on the last day of the month preceding that to which the return refers, and for the annual return the figure should be taken from the entry made on the 31st December of the year previous to that for which the return is being prepared.

6. The figures for the columns headed " Cured," " Relieved," " Discharged otherwise," " Died," should be equal to the sum of the totals given under respective columns similarly named in the Register of In-patients for the period dealt with in the return.

7. The *total number of patients treated both in- and out-door* should be equal to the sum of the figures given in the first four columns of this return.

8. The number of out-patients who *attended personally* and of those *represented by friends* should be obtained from column 2*a* of the Daily Abstract Register. The sum of the figures returned under these headings in the return should be equal to the sum of the figures given as the grand total under the columns headed " Total number of patients treated out-door."

9. The figures of attendance by classes and sexes should be taken from columns 2*b* and 3*a* of the Daily Abstract Register. In grouping the sexes care should be taken to show all under 10 years of age as children, irrespective of sex.

10. The *aggregate of the daily totals of attendance* should be equal to the sum of the totals of columns 2*c* and 3*b* of the Daily Abstract Register.

11. The *daily average attendance for a month* is obtained by dividing the totals of each set of figures given under the column " Aggregate of the daily totals of attendance " by the number of days in the month. The *daily average attendance for a year* is arrived at by dividing the sum of the daily average attendance returned monthly by 12. The daily average attendance should always be represented in whole numbers, particular care being taken to regard remainders in division equal to or greater than the half of the divisor as one; remainders less than the half of the divisor should be thrown out in calculation, thus, supposing the aggregate of the daily attendance to be 529 and the number of days in the month to be 31, the remainder by division, namely, 2, being less than 15½, should be left out in

calculation, and the daily average attendance shown as 17. Supposing, however, the aggregate of the daily attendance in the above case to be 547; the remainder would be 20, which is greater than 15½ (half of 31, the divisor); the daily average attendance should therefore be shown as 18 and not 17. In like manner, supposing the sum of the daily average attendance returned monthly to be 209; the remainder on dividing by 12 would be 5, which, being less than the half of 12, should be omitted and the daily average attendance for the year should be returned as 17. On the other hand, supposing the sum of the daily average attendance returned monthly to be 210, the remainder being 6, or the half of the divisor 12, would cause the daily average attendance for the year to be returned as 18 and not 17.

12. In this return as well as all other returns in which the words "Total treated" occur, the term should be understood to include the number of cases remaining from the previous year or month as the return may happen to be the annual or a monthly return.

XVII.—ANNUAL STATEMENT No. IA.

1. *Column 1.*—The population of the district should be that returned at the last census.

2. *Column 2.*—The classification of dispensaries and hospitals should be drawn up in strict accordance with the instructions contained in Rule 1 of this Manual.

3. *Column 3.*—The year preceding that to which the return relates is referred to.

4. *Column 4.*—The date on which the dispensaries or hospitals shown in column 2 were first established should be given here.

5. *Column 5.*—Should any of the dispensaries or hospitals in a district be either temporarily or permanently closed during the year to which the return relates, the date on which the institution was closed, and, if re-opened, the date on which so opened, should be clearly entered here.

6. *Column 6.*—This column is intended to show the number of dispensaries and hospitals existing in the district at the close of the year to which the return relates.

7. *Column 7.*—The number and date of the authority under which a dispensary or hospital was established, closed, or re-opened, and the reasons for closing the institution should be briefly stated in this column.

XVIII.—ANNUAL STATEMENT No. II.

1. *Column 3.*—The classification is the same as that required under column 2 of Annual Statement No. IA.

2. *Column 4a, b, c, d, and e.*—The figures in these columns should correspond with those given as the total under similar headings in the annual return referred to on pages xxxi to xxxiii and should include those remaining from the previous year as well as those admitted during the year for which the statement is prepared.

e

3. *Column 4f.*—The ratio of deaths *per cent.* of total treated is obtained by multiplying the number under column 4c by 100 and dividing the product so obtained by the number under column 4a; the quotient in this division will be the ratio required. Ratios should be given in whole numbers; the treatment of remainders in calculating averages as explained in clause 11 on pages xxxii and xxxiii should be observed here also.

4. *Column 4g.*—The number of beds which the hospital can accommodate should be a reproduction of the figures given for the same purpose at the head of the annual return. If any change was made in the number of beds during the year to which the statement refers, the increase or decrease in the number, the reasons for the change, and the date from which it was effected, should be all exactly stated in a foot-note.

5. *Columns 4h and 5d.*—The figures given as the daily average attendance of in-patients and out-patients respectively in the statement on the lower portion of the Annual Return (*see* clause 11 on pages xxxii and xxxiii) should be entered in these columns.

6. The figures in *column 5a, b,* and *c* should agree with the figures entered in the first three columns of the statement in the lower portion of the Annual Return (*see* clause 8 on page xxxii).

7. *Column 6.*—The sum of the figures in columns 4a and 5c should be entered here, and should correspond with the number shown as the grand total of the Annual Return (*see* clause 7 on page xxxii).

XIX.—ANNUAL STATEMENT No. III.

1. *Column 3.*—The figures given under the various diseases enumerated under this column should agree with those given against the corresponding diseases under the heading "Total treated, both indoor and out-door" in the annual return (*see* clause 7 on page xxxii).

2. *Column 4.*—The total in this column should agree with the total given under column 6 of Annual Statement No. II.

3. *Column 5.*—The total number of primary and secondary operations performed by Medical Officers and subordinates should be carefully given in accordance with the instructions detailed under Statement No. IIIA (*see* pages xxxv to xxxvii).

4. All Medical Officers and subordinates are enjoined to read carefully pages xxiv to xxx following the preface to the Revised Third Edition of the *Nomenclature of Diseases* issued in 1896.

5. General diseases are no longer divided into groups, as in the last edition of the Nomenclature, and now include certain affections which were formerly returned among local diseases, notably tetanus, Hodgkin's disease (previously returned as lymphadenoma) and diabetes insipidus.

6. Certain parasitic affections such as scabies, phthiriasis, ringworm, &c., formerly returned as parasitic diseases of the skin, are now to be included with other parasitic invasions, and should be returned under column 16 of Annual Statement No. III, "All other diseases" in the division headed "General diseases."

7. The affections known as scrofulous and tuberculous phthisis should be included with diseases returned as "tuberculous

diseases" (*see* column 14 of Annual Statement No. III) and not return-ed under " Other diseases of the respiratory organs."

8. Malarial cachexia and enlargement of the spleen (ague cake) should be returned under " Malarial fever."

9. All diseases, general or local, due to the syphilitic virus, should be returned under " Primary " or " Secondary syphilis ;" so also should " Ulcer of the penis" be returned when due to this cause.

10. In like manner all diseases, general or local, due to the gonor-rhœal virus should be returned under " Gonorrhœa." The following affections will also be so returned when due to this cause :—

> Inflammation of the bladder.
> Inflammation and abscess of the prostate gland.
> Abscess and ulcer of the penis.
> Inflammation and abscess of the spermatic cord.
> Orchitis, acute or chronic, and epididymitis.
> Inflammation of the ovary.
> Inflammation and suppuration of the fallopian tube.
> Inflammation and abscess of the vagina.
> Warts.

11. When the cause of inflammation of the lymph glands is trace-able to syphilis, gonorrhœa, or other specific disease the case should be returned under the heading in which the cause is included.

12. When the cause of any of the functional and symptomatic disorders of the female organs of generation is known, the case should be returned under the appropriate heading.

13. Natural labour is not a diseased condition and should there-fore never appear as such in returns. Should recovery, however, after natural labour be protracted from anæmia or debility, the case should be returned under " Debility and anæmia." The same rule applies to any disease following natural parturition.

XX.—MONTHLY ANNUAL STATEMENT No. IIIA.

1. *Column 1.*—Operations should be classed in strict accordance with the list given on pages 289 to 333 of the Revised Third Edition of the *Nomenclature of Diseases* published in 1896. The classes of operations are indicated in this list by being printed in capital letters, thus :—

> OPERATION ON TUMOURS.
> EVACUATION OF ABSCESSES.
> OPERATIONS FOR ANEURYSM.
> DENTAL OPERATIONS.
> &c.   &c.

There are thirty-three such classes, and on no account should any new class be formed without the previous sanction of the Administra-tive Medical Officer. The classes of operations should be shown in the returns in the same order as given in the Nomenclature.

2. *Column 2.*—The nature of each operation should be described in the same terms as are given under each class in the list referred to above, special care being taken to furnish all necessary infor-

mation in each case, such as stating the method of operation, specifying the site and nature of tumours, abscesses or cysts, foreign bodies, &c., the artery, vein, nerve, bone, joint, &c., operated on.

3. *Column 4.*—The artificial distinction of "Major" and "Minor" operations is abolished and is no longer recognized. Operations are now returned as "Principal" and "Secondary."

(*a*) If only one operation has been performed on a patient, it should be returned as a "Principal" operation.

(*b*) A second operation on the same patient should be also returned as "Principal" if it is in itself distinct and in no way tending to the cure or relief of the condition for which the first was performed.

(*c*) A second operation performed on a patient when the first is surgically cured should be returned separately and classed as a "Principal" operation.

(*d*) All operations tending to the cure or relief of one and the same condition performed on the same or any subsequent day should be returned as one operation.

(*e*) In multiple operations of necessity, such as multiple fractures, dislocations, &c., the main operation should be returned as the "Principal" and the subsidiary operations as "Secondary."

(*f*) When it is considered to be in the interest of the patient that two operations, each in themselves separate and distinct, should be performed at one and the same time, they may be returned as separate operations, each being shown as a "Principal" operation.

(*g*) To secure uniformity in returning dislocations and amputations they should be recorded as follows :—

Operations on joints.
Reduction of dislocation—
Shoulder.
Elbow.
Wrist, &c.
Hip-joint.
Knee, &c.

(*h*) The method of returning amputations in the list given in the *Nomenclature of Diseases* being too complicated for ordinary use, it should be modified as follows :—

Amputations
For injury—
Shoulder.
Arm—Upper third.
Arm—Middle third.
Arm—Lower third.
Elbow.
Fore-arm—Upper third.
Fore-arm—Middle third.
Fore-arm—Lower third.
&c.   &c.
For diseases and deformity, as above.

4. *Columns 3, 5, 6, 7, and 8* refer to individuals operated on and not the number of operations performed on individuals, so that one operation will, as a rule, represent one patient.

5. *Column 9.*—The names and official designations of Medical Officers and subordinates who performed the operations mentioned in the return should be clearly and legibly given under this column.

XXI.—Annual Statement No. IIIB.

1. The name, rank, and grade of each operator should be clearly given.

2. The statement should show all important operations performed at a hospital or dispensary during the calendar year. Officers transferred during the year will not include in the statement they submit any operations performed by them in other districts or at other hospitals or dispensaries.

3. The variety and nature of operations entered in the column "Other operations of importance or interest" should be given in a note at the foot or on the reverse of the statement.

XXII.—Annual Statement No. IV.

1. *Column 3.*—The number of men, women, and children should agree with the figures given as the *total treated* under the heading "Total," "Men, Women, Children" in the statement in the lower portion of the Annual Return (*see* clause 9 on page xxxii).

2. *Column 4.*—The figures given as the total under the various classes in the statement on the lower portion of the Annual Return should be reproduced in this column.

3. Care should be taken to see that the number given under each class is correctly entered, as errors are likely to occur because the order in which the classes are named is not the same in the two statements.

4. *Column 5.*—The *average number* of men, women and children should correspond with the figures given in each of the columns under the heading "Daily average number" of the statement in the Annual Return (*see* clause 11 on page xxxii).

5. The total average number should be equal to the sum of columns 4h, "Total," and 5d of Annual Statement No. II.

6. The *ratio per cent.* is obtained by multiplying the figures in each of the columns under the heading "Average number" by 100, and dividing the product so obtained by the figure entered in the column headed "Total" under "Average number." The sum of the figures given under "Men," "Women," and "Children" should equal 100, never more nor less.

XXIII.—$\frac{\text{Monthly}}{\text{Annual}}$ Statement No. V, Part I.

1. The entries in this statement should include only the receipts and expenditure of the hospital or dispensary proper for the period to which the statement refers. Subscriptions received and amounts expended from the Subscription Fund should not be included in this statement, but should be accounted for separately. The pay of vacci-

nators or other charges on account of vaccination or sanitation, the travelling allowance of subordinates on account of outbreaks of epidemic diseases, attending courts or other places for medico-legal purposes, inspection of villages or outposts, or other charges not relating purely to the dispensary or hospital or its work should not be included in this statement.

Income.

2. *Column A.*—No entry should be made in this column unless a Dispensary or Hospital Fund has been established. The practice of showing here the unexpended budget allotment of the institution is misleading, especially in the annual returns, as the allotments are made for the financial year and the statement is intended to show the income and expenditure of the calendar year.

3. *Column B.*—Particular attention should be paid to the instructions given in notes at the foot of this statement. If Government bears any portion of the cost of maintaining an institution, the amount so paid should be entered under the appropriate headings provided in this column.

4. The amounts shown as " Salaries of Medical Officers " should be the payments made on account of pay and local allowances to officers or subordinates, such as Assistant Surgeons, Hospital Assistants, &c., employed wholly on duties in connection with the hospital or dispensary. When these duties are held as an additional charge, the allowance paid for such charge and no other portion of the Subordinate Medical Officer's salary should be entered in the statement. Allowances given for inspecting animals for slaughter, conservancy arrangements, sanitation, or attending to a lock-up should not be included in the dispensary or hospital accounts.

5. The amount of pay and local or leave allowances charged against a hospital or dispensary should be shown in full. Income-tax or other deductions made on account of loss or damage of hospital property, &c., should not be omitted in this account. House and travelling allowances should be shown under the column referred to in clause 9 below.

6. The salaries paid to all subordinates other than Medical Officers (such as to stewards, matrons, nurses, compounders, ward servants, &c.) should be carefully added together and shown under the heading " Inferior dispensary establishment (including menial servants)." Here also income-tax or other deductions to which the members of this establishment may be subjected should be included.

7. " As registers and forms." Entries under this heading will seldom appear, as under existing arrangements these articles are supplied free of charge.

8. The headings of the next three sub-columns call for no special instruction as they speak for themselves.

9. " Special allowances given by Government." Under this heading should be included all such sums received from Government for purposes which cannot appropriately be classed under any of the other headings in column B. Amounts received on account of build-

ings or repairs, bazaar medicines, travelling and house allowances of subordinates and other miscellaneous charges borne by Government should be entered here.

10. The "total" of the amounts shown under the various headings in column B should be accurately given so as to avoid the statement being returned for correction.

11. *Column C.*—The support received from a Local or a Municipal Fund should be given under appropriate headings, care being taken to include as contributions from Local Funds all amounts received from a District Cess, Town, Port or other fund not included within those of a Municipality and which are raised from local sources.

12. *Columns D and E.*—Entries in these columns will be required only after an institution has been endowed and investments made, securities sold, or deposits withdrawn. These columns should for the present be left blank.

13. *Column F.*—As subscriptions are separately accounted for, no entries should be made in this column.

14. *Column G.*—Amounts realized from paying patients as cost of diets or extras supplied, fees for accommodation, should be shown here.

15. *Column H.*—Amounts received from paying patients and others as value of medicines paid for from funds derived from other sources than Government should be given in this column. Medicines purchased with subscription money should not be included here, as subscriptions are accounted for separately.

16. *Column I.*—The amounts realized on account of sales of instruments, grass or fruit grown within the hospital or dispensary precincts, fines imposed on contractors and others and credited to the income of the institutions should be shown as "Miscellaneous receipts." The sources from which amounts so shown were derived, and the amount obtained from each source, should be distinctly detailed at the foot or on the reverse of the statement.

17. *Column J.*—The "Total income" should be equal to the sum of the amounts entered as the "total" of column B *plus* the amounts detailed under other headings under "Income."

Expenditure.

18. *Column a.*—The expenditure "on establishments" is divided so that the amounts "Paid by Government" and those "Paid from local sources" on this account may be shown separately. When the salaries or any portion of the salaries of the establishment employed at a hospital or dispensary is met from Imperial or Provincial revenues, the amount so expended should be shown as "Paid by Government." Expenditure on account of salaries of establishment met from Municipal, Town, District Cess or other local fund should be shown as "Paid from local sources." Care should be taken to show correctly the distribution of salaries among the various classes of officers and subordinates named in this column.

19. *Columns b, c, d, e, and f.*—The headings of these columns are self-explanatory, but to ensure uniformity in the classification of contingent charges so as to admit of correct comparisons being made, the following instructions should be strictly observed.

20. " *Bazaar medicines.*"—Charges incurred in the purchase of indigenous drugs (*see* Rule 85) and articles used medicinally not ordinarily supplied on indent, *e.g.*, cotton and oil for dressings, cloth for bandages and plasters, rice or wheat-flour (atta), bran, linseed, mustard or other material for poultices, vinegar or rice for lotions, &c.

21. "*European medicines.*"—Drugs, surgical instruments, medical and surgical appliances, and sundries of European manufacture, whether obtained from the Government Medical Stores or local chemists and druggists.

22. "*Diet.*"—Rice, bread, meat, fish, vegetables, milk (fresh or condensed), and other articles prescribed in the sanctioned dietary (*vide* Section VIII, pages xxvi and xxvii), extras, such as aërated waters, eggs, wine, brandy, rum or other stimulant, butter, limes, &c., when not paid for from the Subscription Fund, and such articles as are used in the preparation of diets, as firewood, charcoal, cooking utensils, &c., are to be included as charges under this head.

23. "*Miscellaneous charges.*"—Cloth for sheets, pillow-cases and hospital clothing, straw and coir for mattresses and pillows, blankets, mats, towels, soaps, articles of furniture and repairs to such articles, postal and telegraph or other charges falling under office expenses, railway, boat, cooly and cart-hire, cost of packing cases and all other charges connected with the receipt and despatch of stores, *vide* items 10, 44, 53, 58, 74, 92, and 93 of Appendix E of the Burma Treasury Manual, travelling and house allowances of subordinates and other expenditure which cannot be classified under any of the other headings of the statement. Details of amounts shown in this column should be furnished on the reverse of the statement.

24. "*Buildings or repairs.*"—Construction of, or repairs to, hospital, dispensary or subsidiary buildings.

25. *Column g.*—No entries should be made in this column as none of the existing institutions have been endowed.

26. *Column h.*—The sum of the amounts entered in columns *a* to *f* should be given here. Care should be taken to avoid errors in calculating, as such errors will necessitate statements being sent back for correction.

27. *Cash balance on the last day of the month or year.*—No entry should be made here unless a Hospital or Dispensary Fund has been established (*see* remarks on column A, clause 2, page xxxviii).

28. *The average cost of each diet* should be found by dividing the amount spent on diets shown in column D by the number of diets issued during the period to which the statement refers. The number of diets issued should be ascertained from the Register of diets (*see* page xxvi).

29. *The percentage of total cost paid by Government* is obtained by multiplying the figure returned as the "Total" under column B by 100 and dividing the result so obtained by the "Total expenditure during the month or year" (column *h*). When the entire cost is borne by Government the figure 100 should be entered in this column.

XXIV.—ANNUAL STATEMENT NO. V, PART II.

No returns are made at present in this statement.

1. This statement is required to show the amounts realised as subscriptions and the uses to which such money is put.

2. The receipts during the period to which the statement refers should be carefully given, and the amount given as the opening balance should be taken from the statement for the preceding period ; the amount shown as the closing balance of the period immediately preceding that to which the statement refers should correspond with the opening balance of the succeeding period ; thus the amount given as the opening balance for the month of March should agree with the amount shown as the closing balance of February.

3. The amounts collected should be correctly distributed against the various headings under "Receipts."

4. The *expenditure* should be given in detail on the reverse of the statement, showing how the amounts entered against each of the headings given under "Expenditure classified" has been obtained.

5. Subscriptions are not to be utilized in relieving a hospital or dispensary of any portion of the ordinary cost of maintenance of such institution, and care should be taken to explain distinctly in the column provided for "*Remarks*" why the Subscription Fund has been drawn on for the purchase of any article or other expenditure usually borne, or which would appear to be more correctly borne by the general funds of the institution.

6. The abstract at the foot of this statement is intended for purposes of checking the expenditure. Subscriptions should be expended in proportion to the receipts and not allowed to accumulate.

XXVI.—Subscription Register.

1. In collecting funds from the public, books in Form Medl :—Misc. 16 shall be used. Not more than one register at a time shall ordinarily be used, and, when not in circulation, this register shall be kept in the hospital or dispensary in some convenient place to which the public can have ready access. In the larger towns, where contributors are numerous, more than one book may be found necessary, and under such circumstances care must be taken that the authorized number is not exceeded.

2. Subscribers and donors should be requested to note carefully in the appropriate column the date on which payments are made and to append thereto their initials. Should these particulars be omitted, the Medical Officer in charge or the Secretary should promptly fill in the necessary details.

3. Under the heading "Donations" may be included all such amounts as are received from collection-boxes. The particular source from which such amounts have been derived should be entered in the first column "Name of subscriber or donor."

XXVII.—Medical Case and Certificate for Gazetted Officers.

1. *See* instructions given on form.

2. In cases where they are not actually forbidden in the revised edition of the *Nomenclature of Diseases*, synonyms which are in

f

common use and are desirable on account of their shortness, may be retained when drawing up a medical case, *e.g.*, ague, meningitis, dyspepsia, peritonitis, laryngitis, &c.

3. In order to ensure the accuracy of the terms employed, the Nomenclature numbers of the various diseases should always be given in drawing up medical histories and cases.

XXVIII.—MEDICAL CASE AND CERTIFICATE FOR NON-GAZETTED OFFICERS.

See instructions given on form and clauses 2 and 3 of section XXVII above. Certificates in this form are to be given only when it is desired that leave other than casual leave is to be granted.

XXIX.—MEDICAL CERTIFICATE FOR CASUAL LEAVE.

See instructions on form.

XXX.—MEDICAL CERTIFICATE AND HISTORY OF PERSONS SENT TO LUNATIC ASYLUMS.

See instructions in Appendix F and on form.

XXXI.—LETTER FORWARDING PROPERTY OF DECEASED PATIENTS.

To be sent immediately a patient dies in hospital with all property deposited with the hospital authorities and found on person of the patient.

XXXII.—RECEIPT FOR ARTICLES RECEIVED FOR EXAMINATION.

When vomit, supposed poison or other miscellaneous articles are received, care should be taken to note on the reverse of the form the state in which the articles were received, whether securely packed and sealed, description of seal, and any other information likely to be useful. Articles brought in an insecure condition should not be accepted except in the presence of witnesses, who should attest the correctness of the entries on the reverse of the form with their marks or signatures.

XXXIII.—LETTER ASKING THAT DEPOSITION OF POLICE CASE BE TAKEN.

Requires no special explanation.

XXXIV.—REGISTER OF POLICE CASES.

1. *See* instructions on form.
2. In all cases of "General or Local injuries" care should be taken to specify in the bed-head ticket and the case book, *first*, whether accidental, judicial, homicidal, self-inflicted or other manner in which the injury or injuries were received; *second*, by what mechanism or agent inflicted.

XXXV.—REGISTER OF POST-MORTEM EXAMINATIONS.

All entries should be clearly and legibly made and all corrections and alterations initialled.

XXXVI.—LETTER FORWARDING ARTICLES TO CHEMICAL EXAMINER.

Requires no special explanation.

XXXVII.—Confidential Report on Assistant Surgeons and Hospital Assistants.

To be forwarded to the Administrative Medical Officer not later than the first week after the close of the calendar year, with all particulars carefully and neatly filled in by Civil Surgeon or other Gazetted Officer under whom the subordinate referred to in the report may be serving.

XXXVIII.—Report of making over and assuming charge of Duties.

See instructions on form.

XXXIX.—Transfer Receipt.

See instructions on form.

XL.—Certificate of English Qualification.

See instructions in Appendix A.

XLI.—Results of English Qualification Examination.

See instructions in Appendix A.

XLII.—Nominal-roll of Medical Subordinates desirous of appearing for Professional Examinations.

See instructions on form.

XLIII.—Letter intimating Return, &c., to be blank.

See Rule 196 of Manual.

XLIV.—Quarterly Nominal Disposition Return.

All transfers, new appointments, date of beginning of leave, &c., during the quarter should be given in the column headed " Remarks." This return should be accompanied by a certificate stating that all service books of subordinates named in the return are in the possession of the Civil Surgeon or other officer under whom the subordinates may be serving, and that all entries have been properly made and are up to date.

APPENDIX H.

List of Forms prescribed for use at Civil Hospitals and Dispensaries.

Form No.	Head line of form.	Remarks.
Medical 1	Annual Statement No. 1, showing number of dispensaries and hospitals in the province.	*See* Guard-book of Medical forms.
„ 1A	Annual Statement No. 1A, showing number of dispensaries and hospitals in the district.	*See* p. lxxxvii, App. I.
„ 2	Annual Statement No. II, showing number of in-door and out-door patients treated, &c.	*See* p. lxxxvii, App. I.
„ 3	Annual Statement No. III, showing the $\frac{\text{diseases}}{\text{deaths}}$ of in-door and out-door patients treated, &c. from each of the diseases.	*See* pp. lxxxviii, lxxxix, App. I.
„ 4	$\frac{\text{Annual}}{\text{Monthly}}$ Statement No. IIIA, showing results of surgical operations performed in the $\frac{\text{hospital}}{\text{dispensary}}$, &c.	*See* p. xc, App. I.
„ 5	Annual Statement No. IV showing classes and sexes of in-door and out-door patients treated, &c.	*See* p. xcii, App. I.
„ 6	$\frac{\text{Annual}}{\text{Monthly}}$ Statement No. V, Part I, showing current income and expenditure of hospitals and dispensaries.	*See* pp. xciii, xciv, App. I.
„ 7	Annual Statement No. V, Part II, showing account of invested capital of dispensaries and hospitals, &c.	*See* p. xcv, App. I.
„ 8	Annual Statement No. IIIB.—Selected list of surgical operations.	*See* p. xci, App. I.
„ 9	Bond of a Civil Hospital Assistant ...	*See* pp. xlviii, xlix, App. I.
„ 10	Report of assuming or making over charge ...	*See* Guard-book of Medical forms.
„ 11	Memorandum regarding passing of bill ...	*Ibid.*
„ 12	Memorandum intimating refund	*Ibid.*
„ 13	Proceedings of Medical Board	*See* pp. xcvii to xcix App. I.
„ 14	$\frac{\text{Annual}}{\text{Monthly}}$ return of in- and out-patients treated	*See* pp. lxxx to lxxxvi, App. I.
„ 15	Medical certificate for short leave ...	*See* p. ci, App. I.
„ 16	Statement No. VA.—Subscription account of the $\frac{\text{hospital}}{\text{dispensary}}$ for the $\frac{\text{month}}{\text{year}}$.	*See* pp. xcv, xcvi, App. I.
„ 17	Medical certificate—Non-gazetted officers ...	*See* p. c, App. I.
Medl. Misc. 1	Letter forwarding the property of patients who died at the hospital.	*See* p. ciii, App. I.
„ 2	Charge report	*See* p. cix, App. I.
„ 3	Temperature chart and bed-head ticket combined.	*See* p. liii, App. I.
„ 4	Diet sheet	*See* p. liv, App. I.

Form No.	Head line of form.	Remarks.
$\dfrac{\text{Medl}}{\text{Misc.}}$ 5	Pass from hospital (books of 50 and 200 leaves in counterfoil).	*See* Guard-book of Medical forms.
„ 6	Certificate of discharge from hospital (books of 200 leaves in counterfoil).	*Ibid.*
„ 7	Medical certificate of death (books of 25, 100 and 200 leaves respectively in counterfoil).	*Ibid.*
„ 8	Ticket for out-door patients	*See* p. li, App. I.
„ 9	Daily register of out-patients (200 leaves) ...	*See* p. l, App. I.
„ 10	Daily register of in-door patients (first size 200, second size 100, and third size 50 leaves).	*See* p. lii, App. I.
„ 11	Letter requesting submission of budget estimate.	*See* Guard-book of Medical forms.
„ 12	Daily register of diets	*See* p. lvi, App. I.
„ 13	Daily abstract of in- and out-patients ...	*See* pp. lviii to lxi, App. I.
„ 14	Voucher for bazaar medicines	*See* Guard-book of Medical forms.
„ 15	Voucher for miscellaneous charges ...	*Ibid.*
„ 16	Subscription collection-book	*See* p. xcvi, App. I.
„ 17	Quarterly nominal disposition return of Assistant Surgeons, &c.	*See* pp. cxv, cxvi, App. I.
„ 18	Result of English qualification examinations of Hospital Assistants.	*See* p. cxi, App. I.
„ 19	English qualification certificate for medical subordinates.	*See* p. cx, App. I.
„ 20	Certificate of fitness to return to work (books of 50 leaves in counterfoil).	*See* Guard-book of Medical forms.
„ 21	Receipt for articles sent for medical examination (books of 50 leaves in counterfoil).	*See* p. civ, App. I.
„ 22	Letter to the Magistrate intimating that patient's deposition should be taken at once.	*See* p. cv, App. I.
„ 23	Contingent register (books of 100 leaves and loose sheets).	*See* Guard-book of Medical forms.
„ 24	Letter forwarding articles to Chemical Examiner for analysis.	*See* pp. cvii, cviii, App. I.
„ 25	Report of *post-mortem* examination (books of 50 and 100 forms, and sheets).	*See* pp. cv to cvii, App. I.
„ 26	Statement of breakages and losses of articles...	*See* p. lxxi, App. I.
„ 27	Report of the Chemical Examiner to the local Government upon articles submitted to him for examination or analysis.	*See* Guard-book of Medical forms.
„ 28	Statement showing travelling allowance bills in the .... district cashed during the month.	*Ibid.*
„ 29	Register showing quarterly sales of quinine ...	*Ibid.*
„ 30	Objection statement of items in contingent bill of the $\dfrac{\text{Police}}{\text{Civil}}$ hospital for the month.	*Ibid.*
„ 31	Register of receipts and sales of Government quinine.	*Ibid.*
„ 32	Register of operations performed at the $\dfrac{\text{hospital}}{\text{dispensary}}$.	*See* p. lvii, App. I.

List of Forms prescribed for use at Civil Hospitals and Dispensaries—concluded.

Form No.	Head line of form.	Remarks.
Medl. / Misc. 33	Annual indent for medicines and surgical equipment on Government Medical Stores.	*See* pp. lxii to lxiv, App. I.
„ 33A	Annual indent for medicines and surgical equipment on Messrs. , Wholesale Chemists.	*See* pp. lxvi to lxviii, App. I.
„ 33B	Supplementary and emergent indent for medicines, &c.	*See* pp. lxix, lxx, App. I.
„ 34	Receipt and delivery vouchers	*See* p. lxxii, App. I
„ 35	Surgical equipment ledger	*See* pp. lxxiii to lxxix, App. I.
„ 36	Register of police cases	*See* p. cv, App. I.
„ 37	Confidential report on $\frac{\text{Assistant Surgeons}}{\text{Hospital Assistants}}$	*See* p cviii, cix, App. I.
„ 38	Transfer receipt of medical stores, instruments and appliances.	*See* p. cx, App. I.
„ 39	Nominal-roll of $\frac{\text{Assistant Surgeons}}{\text{Hospital Assistants}}$ desirous of appearing for professional examinations.	*See* p. cxi, App. I.
Misc. / Genl. 93	Report of blank returns	*See* p. cxiii, App. I.

APPENDIX I.

Forms Referred to in the Manual.

FORM OF GUARANTEE.

[See Rule 6.]

KNOW ALL MEN by these presents that we
are held and firmly bound to the SECRETARY OF STATE FOR INDIA
IN COUNCIL in the sum of one thousand rupees to be paid to him his
successors or assigns for which payment to be well and truly made
we bind ourselves and each of us and the heirs executors administrators and representatives of us and each of us and of every of them
jointly and severally firmly by these presents sealed with our seals
dated this   day of   one thousand
hundred and   .

WHEREAS a $\frac{dispensary}{hospital}$ has been established at
                and the Government of Burma has been requested to aid the
same as a $\frac{dispensary}{hospital}$ of the   class pursuant to the revised rules
for the grant of Government aid to charitable hospitals and dispen-
saries in the Province of Burma which the said Government of Burma
has consented to do upon security being given that the local income
from private subscriptions of the said $\frac{dispensary}{hospital}$ shall amount to not less
than   rupees *per mensem* and the above-bounden
                             thereupon agreed to execute the above-
written bond or obligation subject to the condition hereinafter contained : NOW the condition of the above-written bond or obligation is
such that if the local income from private subscriptions of the said
$\frac{dispensary}{hospital}$ shall amount to not less than the sum of   rupees
per mensem or if in case the said local income shall be less than
that sum the above-bounden   or one
of them their or one of their heirs executors administrators or
representatives do and shall monthly and every month pay to the
Committee of Management or Treasurer or other officers or officer of
the said $\frac{dispensary}{hospital}$ for the time being authorized to receive the income
thereof such a sum of money as with the private subscriptions will
cause the said local income to amount to not less than the sum of
        rupees *per mensem* and if the said
or one of them their or one of their heirs executors administrators
or representatives do and shall from time to time and at all times
hereafter save harmless and keep indemnified the said Secretary of
State for India in Council and his successors and assigns and his
and their officers and servants and their and every of their estate
and effects whatsoever from and against all losses and expenses which
shall or may be paid suffered sustained or incurred by him or them
for or by reason or on account of any deficiency or falling off of the
said local income of the said $\frac{dispensary}{hospital}$ whereby it shall be reduced below
the monthly sum of   rupees then the above-written
bond or obligation shall be void otherwise the same shall be and remain in full force and virtue.

PROVIDED that the above-bounden   or one of
them their or one of their heirs executors administrators or representatives desiring at any time hereafter to be released from the
above-written bond or obligation may with the consent of the Government of Burma, withdraw on giving notice of such desire to the Government of Burma through the local Deputy Commissioner or Commissioner and to other parties to this bond (if any) three months be-
forehand.

Signed sealed and delivered by the above-named
in the presence of
Signed sealed and delivered by the above-named
in the presence of
Signed sealed and delivered by the above-named   .
in the presence of
Signed sealed and delivered by the above-named
in the presence of

[*To be executed by Assistant Surgeons and Hospital Assistants
on entering Government Service.*]

FORM $\frac{\text{Medical}}{9.}$

KNOW ALL MEN by these presents that I
son of   of
am held and firmly bound unto the SECRETARY OF
STATE FOR INDIA IN COUNCIL in the penal sum of rupees four hundred
(Rs. 400) to be paid to the said Secretary of State his certain attor-
ney successors or assigns for which payment to be well and truly
made I bind myself my heirs executors administrators and representatives firmly by these presents sealed with my seal dated at
this   day of   .
AND I the said   do hereby for
myself my heirs executors administrators and representatives cove-
nant with the said Secretary of State his successors and assigns that
if any suit shall be brought touching the subject-matter of this bond
or the conditions hereunder written in any Court subject to the High
Court of Judicature at Fort William in Bengal other than the said
High Court in its Original Jurisdiction the same shall and may at the
instance of the said Secretary of State be removed into tried and de-
termined by the said High Court in its Extraordinary Original Jurisdiction.

NOW the condition of the above written bond or obligation which is
executed under the orders of the Government of
and is given for the performance of a public duty in which the public
are interested within the meaning of the exception to section 74 of
Act IX of 1872 is such that if the above-bounden
shall faithfully and diligently serve the said Secretary of .
State his successors and assigns for the period of five years in the
Subordinate Medical Department in the capacity of a Civil Hospital
Assistant and shall and willingly and submissively at all times obey
all the lawful orders and commands of the said Secretary of State his

successors and assigns and of any Commissioned Medical Officer of the department for the time being under whom the above-bounden may be appointed to serve and especially shall when ordered so to do forthwith and as ordered proceed to any part of British India or of its dependencies or of the States in alliance with it where his services shall be required and shall on no account (except in case of illness of which a certificate signed by a Commissioned Medical Officer shall be sufficient proof) without the orders or sanction of the said Secretary of State his successors or assigns leave any station to which he shall have been appointed or in any way terminate his service before the end of the said five years and shall while at every such station well and truly and faithfully perform the duties of a Civil Hospital Assistant and shall also when required so to do and in the event of necessity arising serve on military duty and while engaged on such military duty will strictly conform to and obey and observe all Orders and Regulations prescribed for his conduct subject to the penalties for disobedience therein laid down or failing such to a penalty of Rs. 200 and on any default or breach of any of the provisions hereinbefore mentioned in respect to his duties

   shall well and truly upon demand pay to the said Secretary of State his successors or assigns the sum of rupees two hundred (Rs. 200) for every such default or breach then the above written bond shall be void otherwise the same shall remain in full force and virtue.

Signed in our presence.

g

FORM Medical / 9 Misc.

[See Sections I and IV, Appendix G.]

Register of out-patients treated at the.....................dispensary.

1	2	3	4	5	6	7	8	9	10	11	12	13	14	15	16
Year, month, and date.	Register No. Yearly.	Register No. Daily.	Name or number.	Age.	Class.	Sex.	Occupation.	Residence.	Attended Personally. Old.	Attended Personally. New.	Represented by Friends. Old.	Represented by Friends. New.	Disease.	Treatment.	Remarks.
1st January 1897.	1	1	Ramah	14	H.	M.	Coolie	Lammadaw		1			Colic.	...	...
		2	1234 of 4th December 1896						1		1		Diarrhœa.		
	2	3	1260 of 30th December 1896								1		Fever.		
		4	Bamboo	29	N.	N.	P-on	Botataung		1			Caried tooth.		
	3	5	Soormah	5	H.	C.	Nil	Kemmendine			1		Contusion.		
		6	1269 of 31st December 1896							1		1	Fever.		
	4	7	Rambux	57	H.	M.	Hawker	Lewis street		1		1	Neuralgia.		
	4	7							1	3	2	1			
2nd January 1897.		1	3 of 1st January 1897							1				...	...
	5	2	Nga Ba	44	B.	N.	Broker	Pegu		1			Boil.		
	6	3	Ma Than	14	B.	F.		Alon					Bronchitis.		
		4	1358 of 30th December 1896						1	1					
	7F	5	Roostum	15	H.	M.	Shop-keeper	Yegyaw		1			Wound.		
	3	5							1	3	1				
31st December 1897.	1576	1	M. Brown	37	Eo.	M.	Nil	Strand road		1			Sprain.	...	...
	1577	2	Keshub	43	H.	N.	Clerk	Kokkaing				1	Rheumatism.		
	1578	3	1459 of 23rd December 1897		E.			30th street	1						
	1579	4	Fannie Fuller	10	E.	C.		30th street		1			Mumps.		
		5	Chinee	11	O.C.	F.		China street		1		1	Fever.		
	4	5								3		1		...	...

TICKET FOR OUT-DOOR PATIENTS.

[See Rule 131 and Section II, Appendix G.]

.... *Dispensary.*

Name...

Age....................Class...........................Sex........................

Disease..

Yearly No..

This ticket must be produced by the patient or his friends at all subsequent visits.

Month and date.	Treatment.

Form Medical
10 Misc.

[See Sections III and IV, Appendix G.]

Register of in-patients treated in the ....hospital.

1	2	3	4	5	6	7	8	9	Result.				14	15
Year, month, and date.	Yearly No.	Name.	Age.	Class.	Sex.	Residence.	Occupation.	Disease.	Cured.	Relieved.	Discharged otherwise.	Died.	Remaining.	Remarks.
1st January 1897	1	391 of 1896	43	E.	M.	Aён	Carpenter	Fracture						
1st January 1897	2	450 of 1895	29	O.	F.	Rangoon	Nil	Tumour						
1st January 1897	3	Maung Towa	27	M.	M.	Pegu	Labourer	Cancer						
3rd January 1897	3	Esther	21	E.	M.	Strand road	Nil	Diarrhœa						
4th January 1897		Mahomed												
8th January 1897	4	Henry Heatherwat												
31st January 1897	21	Bakico Perchad	5	H.	C.	Rangoon	Nil	Fever	7	6	4	2	2	Absconded.
1st February 1897	21	2 of 3rd January 1897												
		21 of 31st January 1897	18	E.	F.	Rangoon	Nil	Diarrhœa						
	22	Ma Khin	39	E.	M.	Prome	Tradesman	Wound						
28th February 1897	36	Maung Hpo	25	H.	M.	Aон	Coolie	Fracture						
28th February 1897	401	Budico												
	19								10	2	3	2	3	Removed by friends.
1st December 1897	250	John Williams	36	Eu.	M.	Rangoon	Clerk	Enteric fever						
25th December 1897	265	Daniel O'Connel	21	E.	M.	Rangoon	Nil	Nil						
26th December 1897	266	Mahomed Bakir	65	M.	M.	Do.	Tradesman	Abscess						
29th December 1897	270	Nga Kauk	26	E.	M.	Rangoon	Joiner	Snake-bite						
30th December 1897		Nil					Nil	Nil						
31st December 1897	271	Na Thin	35	E.	F.	Insein	Nil	Debility						
	22								5	4	3	2	8	

FORM $\frac{\text{Medical}}{3 \text{ Misc.}}$

BED-HEAD TICKET AND TEMPERATURE CHART.

[See Rule 119 and Section V, Appendix G.]

Ward No.................Register No.................Medical Officer in charge.................

Name	———	Age———	Class———	Disease or accident———	Result———
Sex	———	Occupation———	Salary———	How long in hospital———	
Religion	———	Abode———		Date of discharge	
Date of admission	———			Date of death —	

Duration of disease.

Date	1	2	3	4	5	6	7	8	9	10	11	12	13	14	15	16	17	18	19	20	21	22	23	24	25	26	27	28	29	30	31
	M E	M E	M E	M E	M E	M E	M E	M E	M E	M E	M E	M E	M E	M E	M E	M E	M E	M E	M E	M E	M E	M E	M E	M E	M E	M E	M E	M E	M E	M E	M E

Temperature
106°
105°
104°
103°
102°
101°
100°
99°
98°
97°

Respiration { M. E.

Pulse { M. E.

Motions ...

Case and symptoms.

Treatment.

Date.

FORM $\frac{\text{Medical}}{4 \ \text{Misc.}}$ DIET SHEET.

[See Rule 182 and Section VI, Appendix G.]

Diet sheet of........................

ADMITTED ON	.	DISEASE.	DISCHARGED ON	DIED ON

Date.	Diet.	Beef.	Chicken.	Mutton.	Eggs.	Custard.	Milk.	Port wine.	Brandy.	Beer.	Butter.	Bread.	Medical Officer's initials.
							EXTRAS.						
1													
2													
3													
4													
5													
6													
7													
8													
9													
10													
11													
12													
13													
14													
15													
16													
17													
18													
19													
20													
21													
22													
23													
24													
25													
26													
27													
28													
29													
30													
31													
Total													

FORM Medical/12 Misc.

DAILY REGISTER OF DIETS.

[See Rule 183 and Sections VII and VIII, Appendix G.]

Daily register of diets issued in the..........................hospital during the month of............................ ....

Date.	1st.	2nd.	3rd.	4th.	5th.	6th.	7th.	8th.	9th.	10th.	&c.
Number of patients dieted ...											
Number of relatives dieted ...											
Number of police cases dieted ..											
Number of patients not dieted.											
Class of Diet.											
Number of full diets ..											
Number of half diets ...											
Number of milk diets ...											
Number of spoon diets											
Number of patients receiving extras.											

Articles.	lbs.	oz.	lbs.	oz.	lbs.	oz.	lbs.	oz.	lbs.	oz.	lbs.	oz.	lbs.	oz.	lbs.	oz.	lbs.	oz.	lbs.	oz.
Rice																				
Dal																				
Fish																				
Beef																				
Oil or ghee																				
Vegetables																				
Salt																				
Condiments																				
Fuel																				
Extras																				

REGISTER OF OPERATIONS.

FORM Medical/32 Misc.

[See Section IX, Appendix G.]

Register of operations performed at the..........................hospital dispensary.

NUMBER.		Class of operation	Nature of operation.	RESULT AND DATE.				Cause of death.	Operator's name and rank.
Serial.	In hospital or dispensary register.			Cured.	Relieved.	Discharged otherwise.	Died.		

Daily Abstract showing class, sex, diseases. &c., of in and out patients treated

Date.	Attended personally.	Represented by friends.	Europeans.			Eurasians.			Burmese.			Mahomedans.			Hindus.			Other classes.			Total of new admissions.			Total of old and new cases treated.			Total.
			M	F	C	M	F	C	M	F	C	M	F	C	M	F	C	M	F	C	M	F	C	M	F	C	

Out-patients.

Class and sex of new admissions.

Remaining from previous month ...																											
1 ...																											
2 ...																											
3 ...																											
4 ...																											
5 ...																											
6 ...																											
7 ...																											
8 ...																											
9 ...																											
10 ...																											
11 ...																											
12 ...																											
13 ...																											
14 ...																											
15 ...																											
16 ...																											
17 ...																											
18 ...																											
19 ...																											
20 ...																											
21 ...																											
22 ...																											
23 ...																											
24 ...																											
25 ...																											
26 ...																											
27 ...																											
28 ...																											
29 ...																											
30 ...																											
31 ...																											
Total admitted ...																											
Remaining at end of the month.																											

at the........... .... $\frac{hospital}{dispensary}$ during...............................

	3		4
	IN-PATIENTS.		POLICE CASES.

	a						b	c	d		
	Class and sex of new admissions.							DIS-CHARGED.			

Europeans.			Eurasians.			Burmese.			Mahomedans.			Hindus.			Other classes.			Total of new admissions.			Total of old and new cases treated.			Total.	Cured.	Relieved.	Otherwise.	Died.	Not dieted.	Admitted.	Discharged.
M	F	C	M	F	C	M	F	C	M	F	C	M	F	C	M	F	C	M	F	C	M	F	C								

h

FORM $\frac{\text{Medical}}{\text{13 Misc.}}$ —concluded.

DAILY ABSTRACT OF

Daily Abstract showing class, sex, diseases, &c., of in and out patients treated

1																							5
	GENERAL DISEASES.																						
Date.	Smallpox.	Cholera.	Dysentery.	Malarial fevers.	Primary syphilis.	Secondary syphilis.	Gonorrhœa.	Scurvy.	Worms.	Debility and anæmia.	Rheumatic affections.	Tubercular diseases.	Leprosy.	All other general diseases.	Diseases of the nervous system.	Diseases of the eye.	Diseases of the ear.	Diseases of the nose.	Diseases of the circulatory system.	Diseases of the lungs.	Other diseases of the respiratory system.	Diarrhœa.	Dyspepsia.	Diseases of the liver.
Remaining from previous month																							
1																							
2																							
3																							
4																							
5																							
6																							
7																							
8																							
9																							
10																							
11																							
12																							
13																							
14																							
15																							
16																							
17																							
18																							
19																							
20																							
21																							
22																							
23																							
24																							
25																							
26																							
27																							
28																							
29																							
30																							
31																							
Total admitted ...																							
Remaining at end of the month.																							

at the............................... $\dfrac{hospital}{dispensary}$ during .... —concld.

LOCAL DISEASES.																OPERATIONS.				
b																Kind.		Results.		
Other diseases of the digestive system.	Diseases of the spleen.	Diseases of the lymphatic system.	Goitre.	Diseases of the urinary system.	Soft chancre.	Other diseases of the generative system.	Diseases of the organs of locomotion.	Diseases of the connective tissue.	Ulcers.	Other diseases of the skin.	All other local diseases.	General injuries.	Local injuries.	Poisons.	Total admitted (new cases).	Number of patients operated on.	Principal.	Secondary.	Cured.	Relieved.	Discharged otherwise.	Died.

Civil Institution
Indent No............

[See Rules 75, 76, 78 and 79,

MEDICAL STORE

Indent on the Medical Store Depôt.............................for medical and surgical
months commencing from

1	2	3	4	5
Consecutive No.	Articles.	Balance on hand on the	DURING YEAR	
			Received.	Expended.
		lb. oz. dr.	lb. oz. dr.	lb. oz. dr.
	I.—MEDICINES ...			

Consecutive No.	Articles.	Balance on hand on the	Since received.	RETURNED TO DEPÔT	
				Repairable.	Unserviceable.
1	2	3	4	5	6
	II.—SURGICAL INSTRUMENTS.				
	III.—MEDICAL AND SURGICAL APPLIANCES.				
	IV.—SUNDRIES.				
	V.—CHEMICAL APPLIANCES.				
	VI.—BOOKS, &c.				

DECLA

I SOLEMNLY declare upon my honour that the medicines, &c., supplied on my
strictly imposed upon me by my public duty.
I also declare that I have carefully ascertained the quantities returned above as
been filled in with my own hand.

Indent approved.

....
Administrative Medical Officer.

Date...................................

NOTE.—Dry articles, mineral acids, chloroform, essential oils, oil terebinth and

SURGICAL EQUIPMENT.

and Section XI, Appendix G.]

DEPARTMENT.

equipment required for the use of the...for twelve
the.........of..............................

AVERAGE DAILY SICK................ ....

6	7	8	9	10	11
			REMARKS BY		Consecutive No.
Remaining in store.	Now required.	Quantity sanctioned.	Administra-tive Medical Officer	Depôt.	
lb. oz. dr.	lb. oz. dr.	lb. oz. dr			

Expended during the year.	Remain in in store.	Now requir-ed.	Number sanctioned.	REMARKS BY		Consecutive No.
				Administra-tive Medical Officer.	Depôt.	
7	8	9	10	11	12	13

RATION.

Indents have been, or shall be, solely expended for the purpose of the service, as

" Remaining in store " to be correct, and that the quantities " Now required " have

....

In medical charge.

carbolic acid are supplied by avoirdupois weight ; fluids by measure. •

FORM $\frac{\text{Medical}}{33 \text{ Misc.}}$ —concluded.

No.

To be submitted in duplicate.

To be forwarded through the Administrative Medical Officer.

Indent for Medical Store Depôt Supplies

required for _____

Hospital for the period ending _____

I.—Medicines.

II.—Surgical Instruments.

III.—Medical and Surgical Appliances.

IV, V, & VI.—Miscellaneous articles.

STATION _____

Dated _____

Register No.

N.B.—1. Medical and other indenting officers are requested to see that full and clear particulars of the address and route by which supplies should be forwarded are in every case carefully filled in in the following form.

2. If supplies are to be forwarded for part of the journey per care of any carrying company or agency, the same should be distinctly stated.

[Here enter full particulars of the address and the route by which supplies should be forwarded.]

Full address ————————————————

Station ————————————————

Route ————————————————

Civil Institution   [See Rules 75, 76, 78, and 79,

Indent No............

Indent on Messrs............... ......Wholesale Chemists,...................for medical

commencing from the....................

1	2	3	4	5
Articles.	Price in sterling money.	Balance on hand on the	DURING YEAR.	
			Received.	Expended.
I.—MEDICINES ...	£ s. d.	lb. oz. dr.	lb. oz. dr.	lb. oz. dr.

Articles.	Price in sterling money.	Balance on hand on the	Since received.	Repairable.	Unserviceable.
1	2	3	4	5	6
II.—SURGICAL INSTRUMENTS.	£ s. d.				
III. -MEDICAL AND SURGICAL APPLIANCES.					
IV.—SUNDRIES.					
V.—BOOKS, &c.					

DECLA

I SOLEMNLY declare upon my honour that the medicines, &c., supplied on my strictly imposed upon me by my public duty.

I also declare that I have carefully ascertained the quantities returned above as been filled in with my own hand.

Indent approved.

....

Administrative Medical Officer.

Date........................ .   .

NOTE.—The above prices are subject to the fluctuations of the market, and supplied only as a guide to All drugs (solid or liquid) are supplied by avoirdupois weight, except where quoted by gallon,
N.B.—When indenting, duty should be estimated on the prices "quoted in bond " at the following spirituous preparations Rs. 6 per gallon ; opium Rs. 12 per pound.
All drugs and chemicals (except quinine, which is free) are now charged 5 per cent. (being import Information as to cost of instruments, &c., not detailed in the above list, supplied on application.

and Section XI, Appendix G.]

equipment required for the use of the...................for a period of..........months
of... .

AVERAGE DAILY SICK....................

6			7			8			9
Remaining in store.			Now required.			Quantity sanctioned.			Remarks by the Administrative Medical Officer.
lb.	oz.	dr.	lb.	oz.	dr.	lb.	oz.	dr.	

Expended during year.	Remaining in store.	Now required.	Number sanctioned.	Remarks by the Administrative Medical Officer.
7	8	9	10	11

RATION.

Indents have been, or shall be, solely expended for the purpose of the service, as

" Remaining in store " to be correct, and that the quantities " Now required " have

....

In medical charge.

cost.
which is 16 fluid ounces to the pound.
rate :—Rectified spirits Rs. 9-6 per gallon ; for sweet spirits of nitre Rs. 8 per gallon, and for all other
duty charge).

FORM Medical/33-A. Misc. —concluded.

No.

To be submitted in duplicate.

To be forwarded through the Administrative Medical Officer.

Indent for Medical Supplies required for _____ Hospital

for the period ending

I.—Medicines.
II.—Surgical Instruments.
III.—Medical and Surgical Appliances.
IV & V.—Miscellaneous articles.

STATION

Dated

Register No.

N.B.—1. Medical and other indenting officers are requested to see that full and clear particulars of the address and route by which supplies should be forwarded are in every case carefully filled in in the following form.

2. If supplies are to be forwarded for part of the journey per care of any carrying company or agency, they should be distinctly stated.

[Here enter full particulars of the address and the route by which supplies should be forwarded.]

Full address

Station

Route

FORM Medical / 338. Misc.

SUPPLEMENTARY AND EMERGENT INDENT FOR MEDICINES, &c.

Indent No....................

[See Rules 77, 78, and 79, and Section XII, Appendix G.]

MEDICAL STORE DEPARTMENT.

Supplementary requisition on the................................ for medicines and other expendible stores required for the use of the................................ hospital for................ months commencing from the................................

1	2	3	4	5	6	7	8	9	10
Consecutive No.	Articles.	Balance on hand on the	DURING YEAR.		Remaining in store.	Now required.	REMARKS BY		Consecutive No.
			Receipts.	Issues.			Administrative Medical Officer.	Depôt.	
		lb. oz. drs.	lb. oz. drs.	lb. oz. drs.	lb. oz. drs.	lb. oz. drs.			

DECLARATION.

I SOLEMNLY declare upon my honour that the medicines and other expendible stores supplied on my indents have been, and shall be, solely expended for the purpose of the service, as strictly imposed upon me by my public duty.

I also declare that I have carefully ascertained the quantities returned above as *Remaining in store* to be correct, and that the quantities *now required* have been filled in with my own hand.

Approved.

Station......................
Date ....

....
In medical charge.

Station....................
Date......................

....
Administrative Medical Officer.

FORM $\frac{\text{Medical}}{\text{33B. Misc.}}$ —concluded.

Supplementary Requisition for Medicines and other expendible stores.

Two copies required.

To be forwarded through the Administrative Medical Officer.

Indent No.

N. B.—1.   Indenting officers are requested to see that full and clear particulars of the address and route by which supplies should be forwarded are in every case carefully filled in below.

2.   If supplies are to be forwarded for part of the journey per care of any carrying company or agency, the same should be distinctly stated.

[Here enter full particulars of the address and the route by which supplies should be forwarded.]

Full address

Station

Route

FORM $\frac{\text{Medical}}{\text{26 Misc.}}$

STATEMENT OF LOSSES, BREAKAGES, &c.

[See Rule 104 and Section XIII, Appendix G.]

Statement showing losses, breakages, and unserviceable articles belonging to the .... $\frac{hospital}{dispensary}$

Serial No.	Articles broken, lost, or otherwise rendered unserviceable.	By whom broken or lost.		How lost, broken, or otherwise rendered unserviceable.	Remarks and recommendation by the Civil Surgeon.	Decision of the Administrative Medical Officer.
		Rank.	Name.			

No.............

FORWARDED through the Civil Surgeon to the Secretary to the Administrative Medical Officer, Burma, for favour of orders.

....

In medical charge of $\frac{\text{Hospital}}{\text{Dispensary}}$.

Dated the ....

Civil Surgeon, ...*District.*

FORM Medical 34 Misc.
No............

RECEIPT AND DELIVERY VOUCHERS.
[See Rules 111, 112, and 113, and Section XIV, Appendix G.]

* ....VOUCHER No............

ALL SERVICES.

I certify that the articles enumerated below have been duly † .... the ‡ ....

and § ....

1		2	3	4				5				6				7
LEDGER		Consecutive No.	Articles.	Quantity.				TOTAL VALUE.								Remarks.
Section.	Page.			lb.	oz.	drs.		Local stores.				Europe stores.				
								Rs.	A.	P.		Rs.	A.	P.		

Note.—The names of the medicines, appliances, instruments, &c., should be entered in column 3 in the same order in which they are arranged in the indent.

Columns 5 and 6 to be filled in by the Examiner, where necessary.

STATION..

Date....................

* Insert " expense," " delivery " or " receipt," as the case may be.

† If " expense," insert *" expended under authority dated* (quote authority by)."

If " delivery," insert *" delivered out of Her Majesty's stores at this place to."*

If " receipt," insert *" received* from / into *Her Majesty's stores by."*

‡ Here state Regiment, Battery, Corps, Department, Hospital, &c.

§ If " expense," insert *" are hereby struck off charge."*

If " delivery " or " receipt," insert *that the issue is in compliance with* (quote the indent or issue order).

Columns 2, 3. and ‡ are the only columns to be filled in when articles are returned into store, the other columns being left blank.

N.B.—The original receipt voucher to be dated, signed, and returned to the issuing officer immediately on receipt of the stores, and the delivery voucher to be retained for submission in the case of officers who are in direct account with the Examiner, Commissariat Accounts, as supporting voucher to the accounts rendered to him.

(Signed)......................

Rank.......   ....

[See Section XV, Appendix G.]

FORM $\frac{\text{Medical}}{35 \text{ Misc.}}$ To be prepared in duplicate.

(*Outer sheet.*)

ONE copy to be forwarded to the Administrative Medical Officer through the Civil Surgeon of the district on the 1st April annually ; the duplicate to be retained on record at the hospital or dispensary to which it relates.

Surgical equipment ledger of the ——————— $\frac{hospital.}{dispensary.}$

Account of equipment in charge of the ... ——— — ——— ———

at the ——————— $\frac{hospital}{dispensary}$.

Showing the quantities received, expended, and otherwise disposed of between the 1st day of April and the 31st day of March——— ———

CONTENTS.

Surgical instruments	... Page		Sundries	... Page
Medical and surgical appliances	,,		Field equipment	... ,,
	Books	Page.		

Daily average sick   Number of beds

.

The attention of officers giving over or assuming charge of surgical equipment invited to Rules 102, 103, 205, and 206 of the Burma Medical Manual.

FORM $\frac{\text{Medical}}{\text{35 Misc.}}$ —continued.

No. and date of voucher.	From or to whom received or issued.							
	RECEIPTS.							
	Remaining on 1st April							
	Total Receipts ...							
	ISSUES.							
	Total Issues ...							
	Remaining on 31st March.							

Remarks as to con-
dition of articles.

k

FORM $\frac{\text{Medical}}{35 \text{ Misc}}$. —*continued.*

No. and date of voucher.	From or to whom received or issued.									
	RECEIPTS.									
	Remaining on 1st April									
	Total Receipts ...									
	ISSUES.									
	Total Issues									
	Remaining on 31st March									

FORM $\dfrac{\text{Medical}}{\text{35 Misc.}}$ —continued.

(Last sheet.) LIST OF VOUCHERS.

List showing number and date of indent in support of receipts recorded.

1.	12.
2.	13.
3.	14.
4.	15.
5.	16.
6.	17.
7.	18.
8.	19.
9.	20.
10.	21.
11.	22.

FORM $\dfrac{\text{Medical}}{35 \text{ Misc.}}$ — *continued.*

List showing number and date of authority for issues recorded.

1.	12.
2.	13.
3.	14.
4.	15.
5.	16.
6.	17.
7.	18.
8.	19.
9.	20.
10.	21.
11.	22.

FORM $\frac{\text{Medical}}{\text{35 Misc.}}$ —*continued.*

Verification statement of equipment by Civil Surgeon.

Articles.	FOUND ON EXAMINATION.						Explanation of Medical officer or subordinate in charge of surplus or deficiency.
	Surplus.			Deficient.			
	Serviceable.	Repairable.	Unserviceable.	Serviceable.	Repairable.	Unserviceable.	

I CERTIFY that the whole of the surgical equipment on charge as entered on pages.................to,, ... have been verified and found to correspond with the entries herein shown, with the exceptions above entered.

I consider the explanations given satisfactory, except in the cases noted over-leaf.

Civil Surgeon.

DISPENSARY

[See Section XVI,

Annual / *Monthly* Return of in- and out-patients treated in the *hospital* / *dispensary* at ....

Class of *hospital* / *dispensory*.

Number of beds available

Diseases.	Remaining from previous *year* / *month*.	Admitted during the *year* / *month*.
	TOTAL NUMBER OF PATIENTS TREATED OUTDOOR.	
GENERAL DISEASES.		
Smallpox		
Cholera		
Dysentery		
Malarial fevers (includes malarial cachexia and ague-cake).		
* Primary syphilis (includes all affections, general or local, due to the syphilitic virus).		
* Secondary syphilis (includes all affections, general or local, due to the syphilitic virus).		
* Gonorrhœa (includes all affections, general or local, due to the gonorrhœal virus).		
Scurvy		
Worms		
Debility and anæmia		
Rheumatic affections (includes rheumatism and myalgia).		
Tuberculous diseases (includes scrofula and tuberculous phthisis).		
Leprosy		
All other general diseases (non-malarial and rheumatic fevers should be included under this heading).		
Total ...		
LOCAL DISEASES.		
Diseases of the nervous system		
Diseases of the eye		
Diseases of the ear		
Diseases of the nose		
Diseases of the circulatory system		
Diseases of the lungs		
Other diseases of the respiratory system		
Carried over ...		

* See note on

.... *for* ....

.... ; Male........., Female.........

IN-PATIENTS.							Total number of patients treated, both in-door and out-door.	Remarks.
Total treated.		Discharged.						
Remaining from previous year / month	Admitted during the year / month	Cured.	Relieved.	Otherwise.	Died.			

next page.

FORM $\frac{\text{Medical}}{14}$ —*continued.*

DISPENSARY

$\frac{Annual}{Monthly}$-*Return of in- and out-patients treated in the* $\frac{hospital}{dispensary}$

Diseases.	Remaining from previous year/month	Admitted during the year/month
	TOTAL NUMBER OF PATIENTS TREATED OUTDOOR.	

LOCAL DISEASES.—concluded.

Diseases.		
Diarrhœa		
Dyspepsia		
Diseases of the liver		
Other diseases of the digestive system		
Diseases of the spleen		
Diseases of the lymphatic system		
Goitre		
Diseases of the urinary system		
* Soft chancre (includes buboes)		
Other diseases of the generative system		
Diseases of the organs of locomotion		
Diseases of the connective tissue		
Ulcers		
Other diseases of the skin		
All other local diseases		
General injuries		
Local injuries		
Poisons		
Total ...		
GRAND TOTAL ...		
Number of cases of natural labour treated ...		

* All venereal diseases will be classed

STATISTICS.

at .... for...............................,—continued.

IN-PATIENTS.						Total number of patients treated, both in-door and out-door.	Remarks.
TOTAL TREATED.		DISCHARGED.					
Remaining from previous $\frac{year}{month}$	Admitted during the $\frac{year}{month}$	Cured.	Relieved.	Otherwise.	Died.		

under one of these four headings.

FORM $\dfrac{\text{Medical}}{14}$—concluded.

Showing the class and sex of the patients treated.

OUT-PATIENTS.			TOTAL TREATED.																				Aggregate of the daily totals of attendance.			Daily average number.			
Attended personally.	Represented by friends.	Total treated.		Europeans.			Eurasians.			Burmese.			Mahomedans.			Hindus.			Other classes.			TOTAL.								
				M.	W.	C.	M.	W.	C.	M.	W.	C.	M.	W.	C.	M.	W.	C.	M.	W.	C.	M.	W.	C.	M.	W.	C.	M.	W.	C.
			In-door ⎰ Remaining f r o m previous $\frac{\text{year}}{\text{month}}$ ⎱ Admitted during the $\frac{\text{year}}{\text{month}}$																											
			Out-door ⎰ Remaining f r o m previous $\frac{\text{year}}{\text{month}}$ ⎱ Admitted during the $\frac{\text{year}}{\text{month}}$																											
			Total																											

NOTE.—All under ten years of age to be reckoned as children.

GENERAL REMARKS.

$\dfrac{\text{ANNUAL}}{\text{MONTHLY}}$ RETURN AND REPORT

OF THE

$\dfrac{\text{Hospital}}{\text{Dispensary}}$ at...

for the $\dfrac{year}{month}$ of...

Received in the office of the Administrative Medical Officer on the

....

N.B.—This return should be despatched as early as possible after the completion of the $\dfrac{\text{year}}{\text{month}}$ to which it refers.

....
In medical charge.

....
Civil Surgeon.

FORM $\frac{\text{Medical}}{\text{1A.}}$

DISPENSARY STATISTICS.

[See Section XVII, Appendix G.]

ANNUAL STATEMENT No. 1A.—*Showing the number and class of* $\frac{hospitals}{dispensaries}$ *in the................district during the year........*

Population of the district.	Class and name of hospitals and dispensaries.	Whether open on the 31st December.	Date on which institution was opened.	If closed during the year, dates from and to which closed.	Whether open on the last day of the year.	Remarks.
1	2	3	4	5	6	7

FORM Medical 2.

DISPENSARY STATISTICS.

[See Section XVIII, Appendix G.]

ANNUAL STATEMENT No. 11.—*Showing the number of in-door and out-door patients treated in the................. hospital dispensary during the year.*

| 1 | 2 | 3 | 4 IN-DOOR PATIENTS. | | | | | | 4 NUMBER OF BEDS AVAILABLE. | | 4 DAILY AVERAGE NUMBER. In-patients. | | | | 5 OUT-DOOR PATIENTS. NUMBER TREATED. | | | | 6 |
Name of district.	Name of hospital or dispensary.	Of what class.	Total treated during the year.	Number cured.	Number relieved.	Discharged, otherwise.	Died.	Ratio of deaths *per cent.* of total treated.	Male.	Female.	Men.	Women.	Children.	Total.	Attended personally.	Represented by friends.	Total treated.	Average daily attendance.	Total number of patients treated, both in-door and out-door.
Total...																			

FORM $\frac{\text{Medical}}{3}$.

DISPENSARY

[See Section XIX,

ANNUAL STATEMENT No. III.—*Showing the diseases of the in-door and out-door*

1	2	GENERAL DISEASES.																	
Name of hospital or dispensary.		Smallpox.	Cholera.	Dysentery.	Malarial fever.	Primary syphilis.	Secondary syphilis.	Gonorrhœa	Scurvy.	Worms.	Debility and anæmia.	Rheumatic affections.	Tuberculous diseases.	Leprosy.	All other diseases.	Diseases of the nervous system.	Diseases of the eye.	Diseases of the ear.	Diseases of the nose.
1	2	3	4	5	6	7	8	9	10	11	12	13	14	15	16	17	18	19	20
Europeans,	In-door ...																		
	Out-door.																		
Natives.	In-door ...																		
	Out-door.																		
Total																			

Column 6—To include malarial cachexia and ague-cake.
Columns 7 and 8—To include all affections, general or
Column 9—To include all affections, general or local,
Column 13—To include (59) rheumatism and (778) myalgia.
Column 14—To include scrofula and (334) Tuberculous
Column 33—To include buboes due to soft chancre.

STATISTICS.

Appendix G,]

patients treated in the $\frac{hospital}{dispensary}$ *of..........................during the year...........*

3																			4	5
LOCAL DISEASES.																				
Diseases of the circulatory system.	Diseases of the lungs.	Other diseases of the respiratory system.	Diarrhœa.	Dyspepsia.	Diseases of the liver.	Other diseases of the digestive system.	Diseases of the spleen.	Other diseases of the lymphatic system.	Goitre.	Diseases of the urinary system.	Soft chancre.	Other diseases of the generative system.	Diseases of the organs of locomotion.	Diseases of the connective tissue.	Ulcers.	Other diseases of the skin.	All other local diseases.	General injuries.	Local injuries.	Poisons.	Total number of in-door and out-door patients treated in each dispensary.	Operations.
21	22	23	24	25	26	27	28	29	30	31	32	33	34	35	36	37	38	39	40	41	42	43

Non-malarial fevers to be entered in column 16.
local, due to syphilitic virus.
due to the gonorrhœal virus. Columns 7, 8, 9, and 32 will therefore include all venereal diseases.
Rheumatic fever to be entered in column 16.
phthisis,

DISPENSARY STATISTICS.

[See Section XX, Appendix G.]

FORM $\frac{\text{Medical}}{4}$.

$\frac{\text{MONTHLY}}{\text{ANNUAL}}$ — STATEMENT No. IIIA.—*List showing the result of all surgical operations performed in the* $\frac{\text{hospital}}{\text{dispensary}}$ *of* .... *during the month of* .... *year*

1	2	3	4			5	6	7				8	9
			NUMBER OF OPERATIONS PERFORMED DURING THE $\frac{\text{year}}{\text{month}}$.					RESULTS OF COLUMN 6.					
Class.	Nature of operations.	Number of patients remaining under treatment on last day of previous $\frac{\text{year}}{\text{month}}$.	Principal.	Secondary.	Total.	Number of patients operated on in column 4.	Total of columns 3 and 5.	Cured.	Relieved.	Discharged otherwise.	Died.	Number of patients remaining under treatment at the close of the $\frac{\text{year}}{\text{month}}$ to which statement refers.	Remarks. (Distinguish operations performed by Medical Officers from those performed by Medical Subordinates.)

FORM $\frac{\text{Medical}}{8}$.

DISPENSARY STATISTICS.

[See Section XXI, Appendix G.]

ANNUAL STATEMENT No. IIIB.—Selected list of surgical operations, showing the number of each performed by individual officers, at the.................... hospital .... dispensary for the year....................

Name of operator.	Period on duty.	Excision of tumours (all kinds).	Excision of joints (partial or complete).	Operations on bones.	AMPUTATIONS.		Extraction of lens.	Abdominal section (for any purpose other than obstetrical).	Abscess of liver (any operation).	REMOVAL OF VESICAL CALCULI.		Obstetric operations.	Other operations of importance or interest.	Total.
					Through and above metacarpus.	Through and above metatarsus.				By incision.	By crushing.			
	M. D.													

FORM Medical / 5.

DISPENSARY STATISTICS.

[See Section XXII, Appendix G.]

ANNUAL STATEMENT No. IV.—*Showing classes and sexes of the in-door and out-door patients treated in the hospital .... during the year .... dispensary .*

1	2	3				4						5							
Name of hospital or dispensary.		TOTAL TREATED DURING THE YEAR.				CLASS.						DAILY ATTENDANCE.							
		Men.	Women.	Children.	Total.	*a* Europeans.	*b* Eurasians.	*c* Hindus.	*d* Mahomedans.	*e* Burmese.	*f* Other classes.	Average number.				Ratio per cent.			
												Men.	Women.	Children.	Total.	Men.	Women.	Children.	Total.
	In-door ...																		
	Out-door ...																		
	Total ...																		

FORM Medical 6

DISPENSARY STATISTICS.

[See Section XXIII, Appendix G.]

MONTHLY / ANNUAL STATEMENT No. V, PART I.—*Showing the current income and expenditure of the* .... *hospital / dispensary*

for the month of .... *year* ....

1	2	3 INCOME																			
		A	B FROM GOVERNMENT.								C FROM LOCAL OR OTHER FUNDS.		D	E	F SUBSCRIPTIONS.		G	H	I	J
			As salaries.																	
Name of district.	Name of hospital or dispensary.	Cash balance on 1st day of the month / year	Medical Officers,*	Inferior dispensary establishment (including menial servants),†	As registers and forms.	As European medicines.	For diet, including police cases.	Sale of medicines supplied by Government.	Special allowances given by Government.	Total.	Local funds.	Municipal funds.	Interest on investments.	Sale of securities or withdrawal of deposits.	From Europeans.	From Natives.	For diet (by paying patients, &c.).	From sale of medicines not supplied by Government.	Miscellaneous receipts.	Total income.
	Total ...																			

* In this current account should be shown (1) fees (*e.*) of Surgeons, Assistant Surgeons, Apothecaries, Hospital Assistants, and Native Doctors *attached* to the dispensary, including special dispensary allowance (if any), and (2) the special allowance given to Surgeons, Assistant Surgeons, Apothecaries, Hospital Assistants, and Native Doctors for holding charge of the dispensary in addition to their ordinary civil work. The pay of the Civil Surgeons or other Medical Officers holding the charge of the dispensary in addition to other duties should not be shown at all in this statement.

† In this column should be shown the salaries of (1) compounders, dressers, &c., (2) menial servants, e.g., cooks, sweepers, chaukidars, wardservants, &c.

FORM Medical —*concluded.*
6

DISPENSARY STATISTICS.

MONTHLY · STATEMENT No. V, PART I.—*Showing the current income and expenditure of the*................ hospital
ANNUAL · dispensary

for the $\frac{month}{year}$ *of*.............".......*concluded.*

	2	4 (EXPENDITURE)																4	5	6	7
1		A — ON ESTABLISHMENT								B	C	D	E	F	G						
		PAID BY GOVERNMENT				PAID FROM LOCAL SOURCES															
		As salaries				On salaries															
		Inferior dispensary establishment (including menial servants).†			Medical Officers.*	Inferior dispensary establishment (including menial servants).†			Medical Officers.*												
Name of district.	Name of hospital or dispensary.	Nursing establishment.‡	Compounders, dressers, &c.	Menial servants.†		Nursing establishment.‡	Compounders, dressers, &c.	Menial servants.†		On bazaar medicines.	On European medicines.	On diet.	On miscellaneous charges (including registers, &c., supplied by Government).	On buildings or repairs.	Invested during the $\frac{month}{year}$.			Total expenditure during the $\frac{month}{year}$.	Cash balance in hand on last day of the $\frac{month}{year}$.	Average cost of each diet.	Percentage of total cost paid by Government.
	Total ...																				

* In these columns should be shown (1) the pay of Surgeons, Assistant Surgeons, Apothecaries, Hospital Assistants, and Native Doctors attached to the dispensary, including special dispensary allowance (if any), and (2) the special allowance given to Surgeons, Assistant Surgeons, Apothecaries, Hospital Assistants, and Native Doctors for holding charge of the dispensary in addition to their ordinary civil work. The pay of the Civil Surgeons or other Medical Officers holding the charge of the dispensary in addition to other duties should not be shown at all in this statement.

† In these columns should be shown the salaries of (1) compounders, dressers, &c., (2) menial servants, e.g., cooks, sweepers, chaukidars, ward-servants, &c.

‡ In these columns should be shown the salaries of trained nurses and midwives.

(*Note.*—Vaccinators should not be shown under dispensary establishments.)

DISPENSARY STATISTICS.

[See Section XXIV, Appendix G.]

STATEMENT No. V, PART II.—*Account of invested capital of the* $\frac{hospital}{dispensary}$

of........ for the year..........

1	2	3	4	5	6	7	8
Number.	Name of hospital or dispensary.	Balance on 1st January.	Invested during the year.	Total.	Securities sold.	Deposits withdrawn.	Balance on 31st December.

The.....................}   *Civil Surgeon.*

FORM $\frac{\text{Medical}}{16}$.   **[See Section XXV, Appendix G.]**

STATEMENT No. VA.—*Subscription account of the............ ....*

$\frac{hospital}{dispensary}$ *for the* $\frac{month\ of........}{year}$ ....

Receipts (particulars).	Amount.	Remarks.
	Rs. A., P.	
Balance on hand on first day of $\frac{month}{year}$ ...		
Subscriptions received from Europeans during the $\frac{month}{year}$		
Subscriptions received from Natives during the $\frac{month}{year}$		
Amount obtained from collection-boxes		
Interest on amount deposited in Post Office Savings Bank.		
Miscellaneous receipts during the $\frac{month}{year}$		
Total ...		

Expenditure classified (for details see reverse) ...	Amount.	Remarks.
	Rs. A., P.	
Nursing establishment		
Menial servants		
Bazaar medicines		
European medicines		
Diet		
Furniture		
Bedding and clothing		
Miscellaneous		
Total ...		

FORM $\frac{\text{Medical}}{16}$—*concluded.*

Rs. A. P.

Cash in hand on 1st January...... ...
Total receipts since 1st January......(including month of account).
Total expenditure on 1st January......(including month of account).

Balance in hand on last day of $\frac{\text{month}}{\text{year}}$ ...

[REVERSE.]

Details of expenditure of subscriptions of the................ $\frac{\text{hospital}}{\text{dispensary}}$ *for the*

month of.........
$\overline{\text{year}}$

Article.	Quantity pur-chased.	Rate.	Amount.	Remarks.
			Rs. A. P.	
		Total ...		

FORM $\frac{\text{Medical}}{16 \text{ Misc.}}$

[See Rule 171, and Section XXVI, Appendix G.]

.... $\frac{\text{hospital}}{\text{dispensary}}$

Hospital subscription register for the month of......................

Name of donor or sub-scriber.	Rate of subscription.	Arrears of subscription.	Amount of subscription paid.	Donation.	Date of payment and initials.	Balance of subscription outstanding.	Initials of Medical Officer.	Date of remittance to the Savings Bank.	Remarks.
	Rs. A.	Rs. A.	Rs. A.	Rs. A.		Rs. A.			

FORM Medical / 13.

[See Rules 163, 164, 165, 168, 169, and Section XXVII, Appendix G.]

PROCEEDINGS OF A BOARD OF MEDICAL OFFICERS ASSEMBLED ON THE..........DAY OF..........

AT..........

TO REPORT ON THE STATE OF HEALTH OF..........

....

President..........

Members {..........

....

WE do hereby certify that, according to the best of our professional judgment, after careful personal examination of the case, we consider the health of..........to be such as to render leave of absence for a period of..........absolutely necessary for his recovery.

..........President.

..........} Members.

....

Instructions.

1. With the cognizance of the head of his office, or, if he is himself the head of the office, of the head of his department, the applicant must present himself with the statement of his case in a cover sealed by his medical attendant at the seat of the Government under which he is serving, or at such other place as may be appointed by that Government where a committee of medical officers can be assembled.

2. If it can be avoided, the medical attendant on a case should not be a member of the Board, and under no circumstances can the medical attendant sit as president. He should, however, attend the Board to furnish any information which may be required if the Board is assembled at the place in which he is residing.

3. Before deciding whether to grant or refuse the certificate, the Board may in a doubtful case detain an applicant under professional observation during a period not exceeding 14 days.

4. After a careful examination of the officer, and consideration of the medical case, the Board will record the period of leave considered necessary.

Care should be taken that leave in excess of the period to which the sick officer is entitled is not recommended.

The period of leave should, as far as possible, be specified in years and months.

Note.—In the case of officers of Her Majesty's Indian Marine, the Board should state—

(i) whether the illness has or has not been, or how far it has been, the result of climate and caused in and by the service;

(ii) whether change of climate in or out of India is necessary.

In urgent cases, when a Medical Board considers it absolutely necessary that an officer should proceed in anticipation of sanction, a recommendation to that effect should be made in the proceedings.

Instructions.

Whenever a Medical Officer considers an officer under his charge to be in such a state of health as to render leave necessary, he should draw up in duplicate a statement of the officer's case in this form, and with the cognizance of the head of the officer's department forward it with the certificate over-leaf duly signed through the head of the sick officer's department with a view to the officer being brought before a Medical Board.

One copy of this statement shall be given to the officer himself for the use of his next medical attendant, the other copy or copies being treated as indicated above.

The medical case should contain a brief but distinct account of the cause of disease, its duration. symptoms, and the treatment adopted.

A clear opinion should be included as to whether or not, or how far, the disability was contracted in and by the service and from the effects of climate, with the reasons for such conclusions.

FORM $\frac{\text{Medical}}{13}$ continued.

This form should be used for all gazetted Civil Officers and all Commissioned Officers under the Civil Leave Rules.

Required in triplicate in cases where leave to Europe or the colonies is recommended ; in all other cases in duplicate.

Statement of case of—

1. Service or department..................................
2. Age...
3. Total service...
4. Service in India...
5. Previous periods of leave { ....
   on medical certificate.
6. Date of return from last leave...
7. Temperament...
8. Habits...
9. Disease...

MEDICAL CASE.

SUMMARY.

FORM $\dfrac{\text{Medical}}{13}$ —*concluded.*

...., Civil Surgeon

I ....

of ...., do hereby certify that ....

of the .... service, is in a bad state of health,

and I solemnly and sincerely declare that, according to the best of

my judgment, a change of air is essentially necessary to his re-

covery, and do therefore recommend that he may be permitted to

proceed to * ....

A Civil Surgeon before granting a medical certificate for leave or extension of leave to an officer absent from his corps or appointment, will, except in case of emergency, require the officer to obtain a statement of his previous medical history.

* Here mention the place to which a change is recommended.

$\left.\begin{array}{l}\textit{Date}...................... \\ \textit{The}...................... \end{array}\right\}$

Civil Surgeon.

FORM $\dfrac{\text{Medical.}}{17}$

[See Rules 163, 166, 168, and 169, and Section XXVIII, Appendix G.]

MEDICAL CERTIFICATE.

ARTICLE 903, CIVIL SERVICE REGULATIONS.

(For non-Gazetted Officers.)

I..

(designation).....................................at...

do hereby certify that..

is in a bad state of health as he is suffering from *.............................

....

....†

and I solemnly and sincerely declare that, according to the best of my judgment, after careful personal examination of the case, I consider the health of.................................to be such as to render leave of absence for a period of .... absolutely necessary for his recovery.

(*Signature*)................................ ....

(*Designation*)........ ....

Dated...............................

COUNTERSIGNED.

(*Signature*)...

(*Designation*)..

Dated...............................

---

* Here state the nature of the illness, its symptoms, causes, and duration.
† If the space here is insufficient, the remarks may be continued on reverse.

FORM $\frac{\text{Medical}}{15}$.

MEDICAL CERTIFICATE FOR CASUAL LEAVE.

[See Rules 167 and 169.]

THIS is to certify that I have personally examined Mr. * ....
.... of the........ .... Department and consider him
to be temporarily incapacitated for the performance of his duties
in consequence of.......................................

His indisposition is *not*,† in my opinion, the result of excess of
any kind,† and I recommend that he should have.................
days' leave for rest and treatment.

(*Signature*)................................

(*Designation*)..........................

Date.....................

N.B.—This certificate should be given only after a personal
examination of the patient and in cases where casual leave is recommended.

* Name and designation, *e.g.*, Clerk, Overseer, &c.
† *Delete* the words in italics when necessary.

FORM $\frac{\text{Medical}}{15}$.

MEDICAL CERTIFICATE FOR CASUAL LEAVE.

[See Rules 167 and 169.]

THIS is to certify that I have personally examined Mr. * ....
.... of the........ .... Department and consider him
to be temporarily incapacitated for the performance of his duties
in consequence of.......................................

His indisposition is *not*,† in my opinion, the result of excess
of *any kind*,† and I recommend that he should have..........
days' leave for rest and treatment.

(*Signature*)................................

(*Designation*)..........................

Date..............

N.B.—This certificate should be given only after a personal
examination of the patient and in cases where casual leave is recommended.

* Name and designation, *e.g.*, Clerk, Overseer, &c.
† *Delete* the words in italics when necessary.

FORM $\frac{\text{Medical}}{17\text{L.A.}}$.

[See Rule 170 and Appendix F.]

CERTIFICATE OF THE MEDICAL OFFICER.

(*See* secs. 4 and 8, Act XXXVI of 1858.)*

The........................ ....

I, the undersigned †..
hereby certify that I, on the.................... day of......................
at.......................... personally examined‡ ....
......................................and that the said..................................
is a § ....and a proper person to be taken
charge of, and detained under care and treatment, and that I have
formed this opinion on the following grounds, namely :—

Facts indicating insanity as observed by myself.

|| ....

Other facts indicating insanity communicated to me by others.

¶ ....

Civil Surgeon.

for MEDICAL HISTORY SHEET,

FORM $\frac{\text{Medical}}{17\text{L. A.}}$—*continued.* 　　　*Please see reverse.*

[REVERSE.]

MEDICAL HISTORY SHEET.

*(If any of the particulars in this statement be not known, the fact is
to be stated.)*

1. Name of patient in full and caste or race.
2. Name of patient's father.
3. Sex and age of patient.
4. Marks whereby the patient may be identified.
5. Married or single or widowed.
6. Condition of life and previous occupation (if any).
7. Religion.
8. Place of birth and recent place of abode.
9. Whether homeless or living with relatives or friends.
10. Previous history and habits.
11. State of bodily health.
12. Whether any member of patient's family has been or is
affected with insanity.

---

* *Note.*—The certificate on this page is required only in the case of *persons* sent
to the Lunatic Asylum under sections 4, 5 or 8 of Act XXXVI of 1858. But the
Medical History sheet on the following pages is required not only in these cases,
but also in all other cases in which a lunatic is sent to the asylum (see Circular
No. 98 of 1896).

† Name and official designation.

‡ Name and residence of lunatic.

§ Lunatic, or an idiot, or a person of unsound mind.

|| Here state the facts.

¶ Here state information and from whom received.

13. Whether the attack is the first attack of insanity or not
14. Age (if known) at onset of first attack.
15. Duration and nature of any previous attacks.
16. Duration of existing attack.
17. Symptoms exhibited.
18. Supposed cause of insanity.
19. Supposed exciting cause of present attack.
20. Whether subject to epilepsy.
21. Whether suicidal.
22. Whether dangerous to others.

Civil Surgeon.

Dated at................

Magistrate or Judge.

The..........................

FORM $\frac{\text{Medical}}{\text{1 Misc.}}$ ·

[See Rule 126.]

No................

FROM

THE CIVIL SURGEON,

....

To

THE DISTRICT SUPERINTENDENT OF POLICE,

....

Dated..................... ....

SIR,

I HAVE the honour to forward herewith, for disposal under section 25 of the Police Act, 1861, the property of patients, as per list given below, who died at this hospital.

I have the honour to be,

SIR,

Your most obedient servant,

Civil Surgeon,
or Medical Officer in charge.

No.	Name.	Property.	Date of death.	History or remarks.

FORM Medical [See Rule 218 and Section XXXII, Appendix G.]
21 Misc

CRIMINAL INVESTIGATION No.............

RECEIVED at......$\frac{\text{A.M.}}{\text{P.M.}}$, on the......... of ...., from

Name................................ Rank.................

Residence........................for examination

and report, the following :—

Dead body.	Clothing	Other substances.
Supposed name	On head	Injuries
Male or female	On trunk	Supposed poison
Apparent age	On limbs	Miscellaneous
Apparent race	Separately	

Receipt granted to (name)..................

(rank)..................

PLACE..................

Dated the........of..........................Medical Officer.

FORM Medical [See Rule 218 and Section XXXII, Appendix G.]
21 Misc

CRIMINAL INVESTIGATION No.............

RECEIVED at......$\frac{\text{A.M.}}{\text{P.M.}}$ on the......... of ...., from

Name................................ Rank.................

Residence........................for examination

and report, the following :—

Dead body.	Clothing	Other substances.
Supposed name	On head	Injuries
Male or female	On trunk	Supposed poison.
Apparent age	On limbs	Miscellaneous.
Apparent race	Separately	

Receipt granted to (name)..................

(rank)..................

PLACE..................

Dated the........of..........................Medical Officer.

FORM $\frac{\text{Medical}}{\text{22 Misc}}$

[See Rule 214.]

MEMO. No...............

To

THE.........................MAGISTRATE OF

....

Dated..

THE undersigned has the honour to report that...........................
admitted into hospital on the.................................(*vide* requisition
No..........from District Superintendent of Police to Civil Surgeon) is
in a dangerous state.

The man is now sensible and his deposition should be taken at
once.

Medical Officer.

FORM $\frac{\text{Medical}}{\text{36 Misc}}$.

[See Rule 213 and Section XXXIV, Appendix G.]

POLICE CASE No...............

1. Name of injured person.
2. Father's name.
3. Occupation of injured person.
4. Age.
5. Sex.
6. Class.
7. Residence.......................village or town...................................district.
8. Nature and extent of injuries.

    (Here enter)—

    (*a*) Size of injuries in inches, length, breadth, and depth.
    (*b*) Part of person on which inflicted.
    (*c*) Whether considered slight, severe, or dangerous.
    (*d*) Kind of weapon by which inflicted and any other remarks which
    may seem desirable to record.

9. Date and hour of admission.
10. Date and hour when first seen by Medical Officer.
11. Detailed history of case and treatment from date of admission to that of
discharge.

FORM $\frac{\text{Medical}}{\text{25 Misc}}$.

[See Rule 240 and Section XXXV, Appendix G.]

FORM FOR POST MORTEM EXAMINATIONS.

Information furnished by Police, or précis of case.

Form $\frac{\text{Medical}}{\text{25 Misc.}}$ —continued.

Corpse identified by...

History of case—

Date and hour of despatch of body—	Date and hour of autopsy—	Name of officer by whom examination was actually made.
Date and hour of receipt		

Appearance of body—
    Muscularity.
    Stout.
    Emaciated.

Special marks—
    Scars.
    Tattooing.
    Amount of hair, &c.

Signs of decomposition—

Wounds and bruises—
    (a) Position.
    (b) Character.
    (c) Size.

State of natural orifices—
    Eyes.
    Ears.
    Nostrils.
    Mouth.
    Vagina.
    Anus.
    Urethra.

State of limbs, &c.—
    Rigor mortis.
    Position.
    Contents of hands, if clenched.
    Features .. { Relaxed. { Contracted.
    Pupils.
    Contents of mouth.
    Position of tongue.
    State of teeth.

Thorax—
    Ribs.
    Cartilages.
    Pleura.
    Pericardium.
    Heart ... { Weight. | Shape and size. | Cavities. | Clots, ante or post mortem. { Muscular structure.
    Vessels ... { Clots. { Aneurism. { Atheroma, &c.
    Lungs ... { Appearance. ; Colour. | Consistence. | Adhesions. ; Weight of right lung. { Weight of left lung.
Larynx, trachea, and broncho for foreign bodies or disease.

FORM $\frac{\text{Medical}}{25 \text{ Misc.}}$ —*concluded.*

Abdomen—
  Peritoneum.
  Peritoneal cavity—contents.
  Liver and gall bladder—form and size, disease or injury, and weight
  Pancreas—disease or injury, and weight.
  Spleen—disease or injury, and weight.
  Kidneys—disease or injury $\begin{cases} \text{weight of right.} \\ \text{weight of left.} \end{cases}$
  Stomach ... $\begin{cases} \text{Size and general appearance.} \\ \text{Appearance of coats.} \\ \text{Contents, appearance, odour, and quantity.} \end{cases}$
  Intestines $\begin{cases} \text{General appearance and contents.} \\ \text{Appearance of coats.} \end{cases}$
  Generative organs—
  Bladder and contents.
  Uterus—appearance, size and contents.
  Vagina—contents.

Head—
  Scalp.
  Bones—disease or injury.
  Membranes.
  Brain—weight, substance and ventricles.
  Base of skull—fractures, caries, extravasation, &c., &c.

The spinal canal need not be examined, unless any indication of disease or injury exists.

Fractures and dislocations—

More detailed description of injury or disease—

Opinion as to cause of death—
  Station

                                        Civil Surgeon or Medical Officer.
Date

FORM $\frac{\text{Medical}}{24 \text{ Misc.}}$.   [See Rules 244 and 258.]

No............   *Dated..................................... ....*
FROM

      THE ....

                    ....

To
      THE CHEMICAL EXAMINER TO GOVERNMENT,
                        ....

SIR,

  I HAVE the honour to advise despatch of the undermentioned substances for chemical analysis and to append a statement of the *post mortem* appearances observed.

                    I have the honour to be,
                        SIR,
                    Your most obedient servant,

              Signature and designation of Medical Officer.

Note.—If the case was seen during life by the Medical Officer, the symptoms observed and treatment adopted should be added as paragraph 2.
  Description of viscera forwarded for examination.

FORM $\frac{\text{Medical}}{24 \text{ Misc.}}$ —concluded.

Mode of packing.	Copy of label attached to bottle.	
Weight of parcel.		Impression of seal.
Mode of despatch.	Date of despatch.	Date of receipt in Chemical Examiner's office.

FORM $\frac{\text{Medical}}{37 \text{ Misc.}}$.

[See clause 8, Appendix A, and Section XXXVII, Appendix G.]

CONFIDENTIAL REPORT ON $\frac{\text{Assistant Surgeon}}{\text{Hospital Assistant}}$ .... FOR THE YEAR .

Questions.		Answers.
(1) Name and grade.	(1)	
(2) Present appointment.	(2)	
(3) (a) How long has he served under you ?	(3)	
(b) How long in present charge.		
(4) State your opinion as to his—	(4)	
(a) Intelligence.		
(b) Tact.		
(c) Manners.		
(d) Capacity for maintaining discipline.		
(e) Professional knowledge,		
(f) General efficiency, and		
(g) Manner in which he has performed his duties generally.		
(5) What do you know of his habits?	(5)	
(6) If subject of report is an Assistant Surgeon, state whether you consider him fit in every respect to hold independent charge of a district or a civil station; if a Hospital Assistant, state whether you consider him fit in every respect to hold independent charge of a hospital.	(6)	
(7) Has he special qualifications fitting him for any special duty ?	(7)	
(8) Have you any further observations to make ?	(8)	
(9) Have you communicated to him a verbatim copy of any unfavourable remarks you have recorded in this report ?	(9)	

....

*Civil Surgeon or Medical Officer
in charge of Jail.*

(Station) ....

Date............................ (District) ....

Form $\frac{\text{Medical}}{37\ \text{Misc.}}$ —*concluded.*

Statement of Appointments held during the year..........

(The entries in this statement should be made by the subordinate himself.)

Dispensary or hospital to which attached.	From	To	Authority under which appointed.

Dated...................... ....

....
Asst. Surgeon or Hospital Asst.

Form $\frac{\text{Medical}}{2\ \text{Misc.}}$.   [See Rules 208 to 210.]

CHARGE REPORT.

No..........., *dated the*.......... ....

(a) Here enter grade or class designation, departmental number, if any, and name.

(a) ....
....

(b) Give the name of the hospital or dispensary or nature of duty engaged on and *station* from which transferred.

on transfer from the (b)
....
relinquished charge of his duties $\frac{\text{before}}{\text{after}}$ noon on the......
.... and

(c) Name of hospital or dispensary, or other duty to which transferred and station. This will be filled in by the officer under whom the subordinate is transferred.

assumed charge at the (c)
....
$\frac{\text{before}}{\text{after}}$ noon on the...............
....

Civil Surgeon,
.....................*District.*

[In cases where a subordinate is transferred out of the district, the lower half of the above report should be left blank and forwarded to the Civil Surgeon or other officer under whom the subordinate is transferred, and the following shall be filled in by the Civil Surgeon from whose district the transfer takes place. A copy of the incomplete report should be sent to the office of the Administrative Medical Officer when the transfer takes place.]

The conduct of the subordinate named above during the period he served under me has been

(d) If unsatisfactory, state reasons on reverse.

(d) ....
....
and I consider him to be

(e) Enter opinion of qualifications, trustworthiness, &c.

(e) ....

....
His service-book has been despatched under registered cover on the .... and all entries in it have been completed.

Civil Surgeon,
or other officer from whose district or supervision the subordinate is transferred.

FORM $\frac{\text{Medical}}{38 \text{ Misc.}}$.

[See Rule 206.]

TRANSFER RECEIPT OF MEDICAL STORES, INSTRUMENTS, AND APPLIANCES.

I HEREBY certify that I have compared the stock of medicines, surgical instruments, and appliances in this $\frac{\text{hospital}}{\text{dispensary}}$ with the ledger and that I find them to be correct, with the following exceptions :—

If there are no differences, the words "with the following exceptions" should be struck out.

....

Relieving Officer.

Accepted.

....

Officer delivering over charge.

FORM $\frac{\text{Medical}}{19 \text{ Misc.}}$.

[See Appendix A, Section ii, clause 17.]

CERTIFICATE.

English qualification certificate for Medical Subordinate of the Hospital Assistant Class.

CERTIFIED that I have this day...................examined No.......
...........grade Hospital Assistant...... .. ..,attached to the .... according to the test laid down in the Medical Manual.

I.—His ability to read ordinary English prose, for instance, a column of the ....is.....................

II.—His knowledge of orthography and ability to write from dictation is ....

III.—His knowledge of simple arithmetic, as far as the Rule of Three, is ....

IV.—His ability to read and write prescriptions in English is......
....

I am of opinion that Hospital Assistant...............................
is qualified and entitled to draw the usual allowance for English qualification.

He has maintained his professional knowledge.

Station .... }
Date...................... }

Civil Surgeon.

FORM $\frac{\text{Medical}}{18 \text{ Misc.}}$ [See Appendix A, Section ii, clauses 17 and 18.]

Result of the English qualification examination of Hospital Assistants...............
who appeared for examination at.............on the..........of.............

No.	Class and name.	Date of rank.	Attached to.	Reading and explanation. 25	Dictation. 25	Arithmetic. 25	Reading and writing prescriptions. 25	REMARKS, (Whether qualified or not.)

....}
The....................... .... }   *Civil Surgeon.*

FORM $\frac{\text{Medical}}{39 \text{ Misc.}}$ [See Appendix A, Section i, clause 10.]

Nominal-roll of $\frac{\text{Assistant Surgeons}}{\text{Hospital Assistants}}$ *serving in the...............District, desirous of*
appearing at the half-yearly professional examination to be held on...............
Dated...........................

1	2	3	4	5	6	7	8	9							10	11
		HOSPITAL ASSISTANT ONLY.						SUBJECTS OF EXAMINATION.								
Grade.	Name.	Departmental No.	Qualified in English.	Language in which questions are to be answered.	Station at which the questions are required.	Date of class.	Date of completion of service in the present grade, rendering applicant eligible for examination.	Surgery.	Anatomy.	Practice of medicine and materia medica.	General sanitation and vaccination.	Medical jurisprudence.	Midwifery.		Date on which promotion is due.	Remarks.

NOTES FOR GUIDANCE.

Column 5—*See* clause 14, Appendix A.
Column 6.—The dates to be entered here should be made in strict accordance with clauses 6 and 7, Appendix A, section i, of the Dispensary Manual. Special concessions should be fully explained in column 11.
Column 9.—If the subordinate has to appear in one or two subjects only, state (in column 11) the date on which he failed in those subjects and give reasons for his not appearing at any intermediate examinations.
Column 10.—Give reasons (in column 11) in case it is desired to have the promotions antedated.
Columns 7 to 10.—Entries to be verified before submission.
Separate forms to be used for Assistant Surgeons and Hospital Assistants. The inappropriate subjects of examination (column 9) for each class should be scored out.
Examinations will ordinarily be conducted at the headquarters station of the district in which the subordinates named in this roll are serving, and, unless the number of subordinates appearing for examination at one time renders it impossible, arrangements for the conduct of their duties should be made locally. When it is not possible to so arrange for duties, the circumstances should be reported at once to the Administrative Medical Officer.

                                        Civil Surgeon,
                          ....District.

....District.

Nominal-roll of Medical Subordinates desirous of appearing at the half-yearly professional examination for promotion.

One copy required.

To be forwarded by the Medical Officer in charge of the district to the Administrative Medical Officer.

FORM $\frac{\text{Miscellaneous}}{93 \text{ General.}}$ [See Rule 196.]

.... DEPARTMENT.

(...............)

No.......

The....................

....

Dated......... ..*the*...........................

MEMO.

THE following returns for the month of...................... being blank, are not forwarded :—

_____

_____

_____

_____

Signature...........................

....

FORM $\frac{\text{Medical}}{\text{17 Misc.}}$

[See Section XLIV, Appendix G.]

....DISTRICT.

Quarterly Nominal Disposition Return of Assistant Surgeons, Hospital Assistants, Medical Pupils, &c., from the 1st of............

to the............of...............

Station................Date.........

Rank and grade.	Name.	Age in years.	Date of present grade.	Length of service in years.	Hospital to which attached at close of quarter.	Conduct and qualifications.	Knowledge of English and date of passing English examination.	Pay and allowances.	REMARKS.—Including removals, postings, leave of absence, and the date to which it extends.

Signature.................

Designation of officer under whom subordinates named are serving.} ....

FORM $\frac{\text{Medical}}{\text{17 Misc.}}$ —*concluded.*

.............................District.

Quarterly Nominal Disposition Return of
Assistant Surgeons, Hospital Assistants,
Medical Pupils, &c., from 1st of.........
.... to the.........of.....................

Station........................

Dated...........................

APPENDIX J.

[See Rule 176.]

Extract of Articles 29, 30, and 31, Burma Account Manual.

ACCOUNTS TO BE KEPT FOR HOSPITAL AND SUBSCRIPTION FUNDS.

29. In Municipalities where a separate Hospital fund is maintained under Circular No. 4 of 1898, the Hospital fund is a part of the Municipal fund, and the procedure laid down in Articles 23 to 28 for School funds should be followed. hospital accounts being treated in a precisely similar manner to School funds, including the periodical crediting of the annual assignments, the preparation of the budget estimates, and the keeping of separate accounts and balances.

30. Subscriptions received from the public for special purposes need not be paid into the treasury, but may be retained by the Medical Officer in charge of the dispensary on his own responsibility, or deposited in the Post Office Savings Bank. He should maintain a simple cash-book for them (Ex. L. F. Form No. 2), showing on the receipt side any opening balance he may have in hand and the sums received during the month, and on the expenditure side his disbursements and the closing balance. On the last day of the month he should balance the cash-book and, if there is a separate Hospital fund, should send the cash-book together with any vouchers for sums paid in excess of Rs. 10 to be incorporated in the Hospital fund accounts, the receipts, expenditure, and balances being shown as part of these accounts and a certificate of their having been incorporated being given in the cash-book itself before it is returned to the Medical Officer. A note should also be made in it of the amount at credit in the Savings Bank.

30A. Beyond seeing that the vouchers for sums in excess of Rs. 10 are forthcoming for the amounts paid and are properly receipted, the auditing officer will not question the nature of the payments of the Medical Officer in charge from the Subscription fund. The auditor should, however, see that subscriptions are duly brought to account and that the charges do not include gratuities or payment of any kind to the hospital staff. This rule applies to all Subscription funds, whether a separate Hospital fund is maintained or not.

31. Where there is a separate Hospital fund, the opening and closing balances will be shown as under in the Hospital fund cash-book maintained at the Municipal office :—

	Rs.
Balance in treasury	175
Balance with Medical Officers as per special Subscription cash-book	25
Total ...	200

INDEX.

C.

W.